•••••••• Understanding Garbage and Our Environment

● Other McGraw-Hill Books by Terrific Science Press

Science Projects for Holidays throughout the Year

Exploring Energy with TOYS

Investigating Solids, Liquids, and Gases with TOYS

Exploring Matter with TOYS

Teaching Physical Science through Children's Literature

Teaching Chemistry with TOYS

Teaching Physics with TOYS

Understanding Garbage and Our Environment

Andrea J. Nolan

Editor
Mickey Sarquis, Director
Center for Chemical Education
Miami University Middletown

Terrific Science Press
Miami University Middletown
Middletown, Ohio

**LEARNING
TRIANGLE
PRESS**

*Connecting kids, parents, and teachers
through learning*

An imprint of McGraw-Hill

New York   San Francisco   Washington, D.C.   Auckland   Bogotá
Caracas   Lisbon   London   Madrid   Mexico City   Milan
Montreal   New Delhi   San Juan   Singapore
Sydney   Tokyo   Toronto

Terrific Science Press
Miami University Middletown
4200 East University Blvd.
Middletown, Ohio 45042
513/727-3318
cce@muohio.edu

Published by McGraw-Hill.

McGraw-Hill

A Division of The **McGraw·Hill** Companies

1 2 3 4 5 6 7 8 9   MAL/MAL   9 0 3 2 1 0 9 8

ISBN 0-07-064760-7

This monograph was developed in partial fulfillment of the Master of Environmental Science degree, Institute of
Environmental Sciences, Miami University. This material is based upon work supported by the Ohio Board of
Regents (grant numbers 5-42, 1-46, and 0-36), the National Science Foundation (grant numbers TPE-8953362 and
TPE-9153930), and the National Institute of Environmental Health Sciences (grant number IR25ESO8192). Any
opinions, findings, and conclusions or recommendations expressed in this material are those of the authors and
do not necessarily reflect the views of the Ohio Board of Regents, the National Science Foundation, or the
National Institute of Environmental Health Sciences.

Library of Congress Cataloging-in-Publication Data applied for

Contents

good

OK – don't know if they got point

✓

✓

● Acknowledgments

The author and editor wish to thank the following individuals who have contributed to the development of this book.

Contributor
Christine Cowdrey, Miami University Middletown, Middletown, OH

Terrific Science Press Design and Production Team
Document Production Manager: Susan Gertz
Technical Coordinator: Lisa Taylor
Technical Writing: Christine Cowdrey, Susan Gertz, Lisa Taylor
Technical Editing: Lisa Taylor, Jenny Stencil
Illustration: Stephen Gentle, Alesia Merris, Susan Gertz, Jenny Stencil
Design/Layout: Susan Gertz
Production: Christine Cowdrey, Becky Franklin, Stephen Gentle, Jennifer Pia, Tracy Scobba, Jenny Stencil, Lisa Taylor

Reviewers
Sandra A. Cook, Ferris State University, Big Rapids, MI
Daniel P. de Regnier, Ferris State University, Big Rapids, MI
Bob Hunt, Franklin Associates, Inc., Prairie Village, KS
Baird Lloyd, Miami University Middletown, Middletown, OH
Paul McConocha, Federated Department Stores, Inc., Cincinnati, OH
Chris Myers, Miami University, Oxford, OH
Tina Selden, Hamilton County Environmental Services, Cincinnati, OH
Jerry Sarquis, Miami University, Oxford, OH
Carolyn Watkins, Ohio EPA, Division of Solid and Infectious Waste Management, Columbus, OH
Rita Voltmer, Miami University, Oxford, OH
Linda Woodward, Miami University Middletown, Middletown, OH

Development Team
Bruce Peters, Project Coordinator, Middletown, OH
Linda Jester, Instructor, Middletown, OH

Ralph Austerman, Middletown, OH
Deanna Brewer, Carlisle, OH
Elaine Fagan, Middletown, OH
Mary Garrett, Middletown, OH
Anne M. Holbrook, Cincinnati, OH
Marsha Myers, Cincinnati, OH

Regina Oglesbee, Centerville, OH
Melissa Pence, Middletown, OH
Lesley T. Plymesser, West Union, OH
Sandy Van Natta, Cincinnati, OH
Connie Wyatt, Cincinnati, OH
Virgina Wysong, Lewisburg, OH

● Foreword

Garbage is something we see every day, handle every day, and produce every day. But how much do we know about where the trash bags go after the garbage collector takes them away? This question goes beyond idle curiosity. Solid waste management is a topic of debate in many communities, for economic, environmental, and health reasons. Making informed decisions on this issue may not be easy, however, as myths about garbage abound and personal interests often conflict with each other.

It is our hope that this module will help students learn that managing garbage involves far more than taking the bags to the curb, which is the extent of most people's contact with solid waste. Solid waste management involves science, technology, politics—even personal values. The activities in this book will help students become aware of this complexity and begin to develop their own ideas.

The purpose of this book is to equip students to make wise choices about solid waste management—and to understand the problems associated with poor choices. The investigations in this book are based on science. They are not designed to produce results that lead students to draw a particular conclusion. Instead, this book is designed to help students determine facts—facts that in some cases will vary from community to community—and stimulate students to think about the implications of those facts for solid waste management. The scientific investigations themselves fit into the context of an imaginary town's decision-making process, for which students must gather evidence throughout the unit. Thus, students will apply their new knowledge in the context of a decision-making process.

The activities in this book have been tested and revised based on the recommendations of hundreds of teachers who participated in various CCE workshops. The goal of these workshops was to educate teachers about environmental health-risk issues and how to make informed personal and social choices about the environment. The text has also been extensively reviewed by experts in waste management.

We hope you find this book to be a useful tool in teaching about the issue of solid waste. You and your students may find garbage more fascinating and surprising than you ever imagined!

Mickey Sarquis, Director
Center for Chemical Education

● Introduction

● ● ● ● ● ● ● ● ● ● ● ●   *Understanding Garbage and Our Environment* provides eight multi-activity
lessons that enable middle-level students to explore important solid-waste
management issues. Understanding the benefits of and risks posed by different
waste management approaches depends on understanding the science,
technology, and economics of each. As consumers and future voters, students in
the middle grades need to realize that their personal consumer choices and the
choices of their communities regarding waste management can affect both
environmental and human health. They need to learn that emotion and hearsay
are not good tools for making the best choices about waste management, but
instead learn to depend on reliable information viewed with a critical eye.
Through the information, demonstrations, and activities in this book, students
will have an opportunity to do so. The key concepts explored in this book are as
follows:

- The environmental and human health of a community depends on effective
  waste management.
- Waste management begins with reduction of waste through minimization of
  materials used by industries and conservation by individuals.
- Waste materials can be diverted from the waste stream and reused by the
  original user or others.
- Waste materials can undergo resource recovery when recycled, composted,
  or used as energy sources.
- Modern landfills are an important part of integrated waste management.
- The challenge of effective waste management has no single solution.

Understanding Garbage and Our Environment can supplement any existing
science, health, or environmental education program while effectively
addressing solid-waste management issues through a) improved knowledge of
basic science and technology, b) improved understanding of scientific inquiry
and the limitations associated with scientific evidence, and c) guided
experiences requiring students to use scientific concepts, processes, and
evidence to make personal and social choices regarding solid waste management.

As much as possible, the investigations included in this book require only
inexpensive materials that are readily available in local hardware, grocery, and
toy stores or that can be collected within the school building or from the
students' homes. Where applicable, teachers are provided with vendor
information for specialized materials and with information on inexpensive
alternatives to more sophisticated laboratory equipment.

● Annotated List of Lessons

● ● ● ● ● ● ● ● ● ● ●

Understanding Garbage and Our Environment is organized into eight lessons, each containing one or more student activities. Ideally, your students should complete at least one activity from each lesson, following the lessons in numerical order. The eight lessons are outlined here:

Lesson 1

Students are introduced to the topic of solid waste management with a town meeting role-playing scenario. In **"Trash Trouble in Tacktown, Part A"** students are introduced to the scenario of a town faced with a solid-waste disposal issue. They begin by generating ideas and are assigned the roles of specific town members. The rest of the lessons in this book equip students with the knowledge and ideas they will need for Part B of the scenario, which is in Lesson 8. The second activity in this lesson, **"Risky Business,"** introduces students to the basic concepts of probability and risk they will need in order to evaluate the dangers of different waste disposal methods. *The Garbage Gazette* **"Getting a Degree in Garbage"** introduces students to the many people who deal with solid waste and the jobs they do.

Lesson 2

Students investigate the principle of waste characterization through **"What's Waste?,"** a hands-on inventory of classroom waste. The results of the trash inventory are graphed and compared to nationwide averages. *The Garbage Gazette* **"Garbage Takes a Vacation"** retells the story of the *Mobro 4000,* the notorious garbage barge that couldn't find a place to unload in 1987. In a second *Garbage Gazette*, **"Archaeologists Find Garbage Tells the Truth,"** students are introduced to the people who study our garbage after it's thrown away and the lessons it can teach us about our lifestyles.

Lesson 3

The connection between solid waste disposal and human health is introduced by **"You *Can* Catch Me, You Dirty Rat,"** a contagion simulation that demonstrates to students the communicability of diseases and the concept of exponential growth. Students further explore the topic of diseases and their links to waste management by playing **"Name That Disease,"** a game similar to the popular "Guess Who?" game. *The Garbage Gazette* **"A Thousand Years Without a Bath"** presents changing attitudes about hygiene throughout history and epidemics caused by poor hygiene. **"Toxic Waste and You,"** the second *Garbage Gazette,* explores the story of Love Canal and the issues involved in health research.

Lesson 4

The concept of waste reduction is investigated through two laboratory activities. The first, "**One Liter—To Go**," is a comparative packaging activity in which the masses of different types of beverage containers are calculated and compared. The second activity, "**Wrap It Up**," gives students the chance to study the way bubble gum is wrapped and then challenges them to design their own minimal packaging for the gum. Students are asked to consider the importance of packaging in source reduction. *The Garbage Gazettes* explore two problems of packaging. "**War of the Packing Fillers**" weighs the pros and cons of polystyrene versus starch loosefill packaging "peanuts." "**Enjoying Fresh Milk...In the Desert?**" traces the history of food preservation and explains the pros and cons of aseptic packaging for food items.

Lesson 5

The idea of reusing waste is addressed through two activities which involve making useful products from "waste." The first activity, "**Treasures from Trash**," allows students to create their own inventions and art from used plastic bottles, disposable plates, milk cartons, and other typically discarded items. The second, "**Shrinking Crafts**," shows students how they can reuse polystyrene to make their own shrinking craft key chains, window ornaments, or other items. Students learn about people who reuse materials on a grander scale in *The Garbage Gazette* "**Earthships Take Off**," a description of houses that are built using old tires, pop cans, glass bottles, and other already-used materials. *The Garbage Gazette* "**Don't Dismiss Disposables**" looks at the debate about plates, cups, "to-go" packaging, and other food-service items in restaurants.

Lesson 6

Resource recovery is a broad, multifaceted area consisting of three major categories: composting, recycling, and incineration. Students explore each topic in one or two laboratory activities as well as additional information sheets that can be used as handouts or overheads.

The investigation of composting consists of two laboratory activities. In the first activity, "**Not Eggsactly Decomposing**," students determine the decomposition rates of several different types of solid waste by putting the waste into the wells of an egg carton with soil and water and making long-term observations. In "**Compost Columns**," students create their own compost columns out of plastic 2-L bottles and watch decomposition in action.

The discussion of recycling begins with a review of solid waste characterization and the kinds of items that are recyclable. "**Now Separate It!**," a laboratory activity that uses the specific properties of different types of solid waste to separate them, allows students to explore some of the problems associated with recycling center separation. The second activity, "**Trash in the Newspaper**," allows students to recycle used paper into new paper. During the activity, student groups are challenged to devise removal methods for different types of paper contaminants.

The topic of incineration is covered in the activity "**How Good Is Your Fuel?**" in which students calculate the potential energy value of different items found in the waste stream. *The Garbage Gazette* "**Fluff Up a Milk Jug for a Good Night's Sleep?**" looks at what products are made from recycled materials. The use of worms in composting, a growing trend, is explored in *The Garbage Gazette* "**Nature's Garbage Disposal**."

Lesson 7

To study landfills, the students draw up designs for imaginary landfills in the activity "**Design Your Own Landfill**." In the second activity, "**Believe It Can Rot—Or Not**," students participate in a mock game show that demonstrates the difference in decomposition rates between items buried in a landfill and those exposed to nature's elements. In "**The Bottom Line(r)**," students explore the ability of various materials to contain or slow the movement of water through a landfill.

In the last activity, "**Household Hazardous Waste**," students examine the labels of hazardous products and conduct inventories of hazardous products in their homes. Students take a look at "**A Timeline of Trash Disposal**" in *The Garbage Gazette* included with this lesson.

Lesson 8

In this lesson, which concludes the unit, students study the overall environmental impact of materials in the activity "**Life Cycle Analysis**." Then, students use information they have gathered through their investigation of solid waste management to enact the town meeting in "**Trash Trouble in Tacktown, Part B**." After considering the testimony of the various citizens, the city council comes to a decision on what to do with Tacktown's solid waste. By participating in this role-playing, students realize many factors must be considered when addressing the solid waste problem and no easy solutions exist. Students will have an opportunity to use what they have learned as they look at commonly held myths about solid waste in *The Garbage Gazette* "**Garbage Fact and Fiction**."

● Components of Lessons

● ● ● ● ● ● ● ● ● ● ● Each lesson has several different components, some written to you, the teacher, and others written directly to the student. We suggest that you skim the entire lesson first to get a feel for the topic and then carefully read each of the lesson components to understand them in detail. The following paragraphs provide a brief overview of the lesson components, presented here in the same order as you will see in the lessons. Note that not all lessons contain every component.

Teacher Background provides you, the teacher, with an introduction to (or a review of) the main topic(s) in the lesson. In some cases, one Teacher Background is provided for the entire lesson. In other cases, individual activities within the lesson have their own Teacher Backgrounds. The teacher background material is designed to provide you with information at a level beyond what you will present to your students. You can then evaluate how to adjust the content presentation for your own students.

Teacher Notes gives you detailed information for conducting the activities in your classroom. Notes and safety precautions are included in activities as needed. Teacher Notes consists of up to ten sections, as described here. (The sections listed in brackets may not appear in all Teacher Notes.)

- *Materials:* Materials are listed for each part of the activity, divided into amounts per class, per group, and per student.
- *Safety and Disposal:* Special safety and/or disposal procedures are listed if required.
- [*Getting Ready*]: Information is provided in Getting Ready when preparation is needed before beginning the activity with the students.
- *Opening Strategy:* The Opening Strategy provides a suggested introduction to the student activity.
- [*Procedure*]: Outlines steps for a class demonstration or teacher-led activity.
- [*Discussion*]: Post-activity discussion questions are suggested.
- [*Extension(s)*]: Extension activities for the topic are provided.
- [*Explanation*]: An explanation is provided when information beyond the Teacher Background is needed.
- *Cross-Curricular Integration:* Cross-Curricular Integration provides suggestions for integrating the science activity with other areas of the curriculum.
- [*Resources*]: Resources are listed when the activity is adapted from another source or when a significant amount of information from another source is used. (Other references are listed at the end of the book.)

Student Background provides students with information they will need to be familiar with *before* beginning an activity.

Student Instructions consists of three sections written for the student. The Procedure section gives complete step-by-step instructions for the activity. The Data Recording section provides a place for students to record the results of their investigation. The Analysis Questions ask students to interpret and comment on the results of their investigations. Notes and safety precautions are included in activities as needed and are indicated by the following icons and type style:

   Notes are preceded by an arrow and appear in italics.

   Cautions are preceded by an exclamation point and appear in italics.

Student Information provides students with information they will need to use *during* an activity and can be used as handouts or overheads.

The Garbage Gazettes are additional information resources for students. Written in a fun, newsy style, *The Garbage Gazettes* cover topics related to the main topic of the lesson. One or more *Garbage Gazette* stories are included at the end of each lesson.

● Safety Procedures

● ● ● ● ● ● ● ● ● ● ● ● Experiments, demonstrations, and hands-on activities add relevance, fun, and
excitement to science education at any level. However, even the simplest
activity can become dangerous when the proper safety precautions are ignored
or when the activity is done incorrectly or performed by students without
proper supervision. While the activities in this book include cautions, warnings,
and safety reminders from sources believed to be reliable, and while the text
has been extensively reviewed, it is your responsibility to develop and follow
procedures for the safe execution of any activity you choose to do. You are also
responsible for the safe handling, use, and disposal of chemicals in accordance
with local and state regulations and requirements.

Safety First

- Read and follow the American Chemical Society Minimum Safety Guidelines
  for Chemical Demonstrations on the following page. Remember that you are
  a role model for your students—your attention to safety will help them
  develop good safety habits while assuring that everyone has fun with these
  activities.

- Read each activity carefully and observe all safety precautions and disposal
  procedures. Determine and follow all local and state regulations and
  requirements.

- Never attempt an activity if you are unfamiliar or uncomfortable with the
  procedures or materials involved. Consult a high school or college chemistry
  teacher or an industrial chemist for advice or ask that person to perform the
  activity for your class. These people are often delighted to help.

- Always practice activities yourself before using them with your class. This is
  the only way to become thoroughly familiar with an activity, and familiarity
  will help prevent potentially hazardous (or merely embarrassing) mishaps. In
  addition, you may find variations that will make the activity more
  meaningful to your students.

- Do not conduct activities at grade levels beyond those recommended
  without careful consideration of safety, classroom management, and the
  need for additional adult supervision.

- You, your assistants, and any students participating in the preparation for or
  performance of the activity must wear safety goggles if indicated in the
  activity and at any other time you deem necessary.

- Special safety instructions are not given for everyday classroom materials being used in a typical manner. Use common sense when working with hot, sharp, or breakable objects. Keep tables or desks covered to avoid stains. Keep spills cleaned up to avoid falls.

- When an activity requires students to smell a substance, instruct them to smell the substance as follows: Hold the container approximately 6 inches from the nose and, using the free hand, gently waft the air above the open container toward the nose. Never smell an unknown substance by placing it directly under the nose. (See figure.)

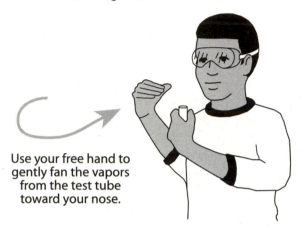

Use your free hand to gently fan the vapors from the test tube toward your nose.

Wafting procedure—Carefully fan the air above the open container toward your nose. Avoid hitting the container in the process.

American Chemical Society Minimum Safety Guidelines for Chemical Demonstrations

This section outlines safety procedures that Chemical Demonstrators must follow at all times.

1. Know the properties of the chemicals and the chemical reactions involved in all demonstrations presented.

2. Comply with all local rules and regulations.

3. Wear appropriate eye protection for all chemical demonstrations.

4. Warn the members of the audience to cover their ears whenever a loud noise is anticipated.

5. Plan the demonstration so that harmful quantities of noxious gases (e.g., NO_2, SO_2, H_2S) do not enter the local air supply.

6. Provide safety shield protection wherever there is the slightest possibility that a container, its fragments, or its contents could be propelled with sufficient force to cause personal injury.

7. Arrange to have a fire extinguisher at hand whenever the slightest possibility for fire exists.

8. Do not taste or encourage spectators to taste any non-food substance.

9. Never use demonstrations in which parts of the human body are placed in danger (such as placing dry ice in the mouth or dipping hands into liquid nitrogen).

10. Do not use "open" containers of volatile, toxic substances (e.g., benzene, CCl_4, CS_2, formaldehyde) without adequate ventilation as provided by fume hoods.

11. Provide written procedure, hazard, and disposal information for each demonstration whenever the audience is encouraged to repeat the demonstration.

12. Arrange for appropriate waste containers for and subsequent disposal of materials harmful to the environment.

● Relationship to National Science Education Standards

Many of the *National Science Education Standards* are supported by the eight lessons in this book. Overall, these lessons provide opportunities for scientific investigation and data analysis in the context of technological applications and impact on society.

As a whole, this book supports the following content standards for grades 5–8:

Science as Inquiry
Content Standard A: As a result of activities in grades 5–8, all students should develop

- abilities necessary to do scientific inquiry, and
- understandings about scientific inquiry.

Physical Science
Content Standard B: As a result of activities in grades 5–8, all students should develop understanding of

- properties and changes of properties in material, and
- transfer of energy.

Science and Technology
Content Standard E: As a result of activities in grades 5–8, all students should develop

- abilities of technological design, and
- understandings about science and technology.

Science in Personal and Social Perspectives
Content Standard F: As a result of activities in grades 5–8, all students should develop understanding of

- personal health;
- populations, resources, and environments;
- risks and benefits; and
- science and technology in society.

● Involvement of Solid Waste and Environmental Professionals

While the lessons in this book provide an overview of the impact solid waste has on us and our environment, at times you may wish to enhance coverage of the topic with real-world examples and perspectives. Inviting professionals from the fields of environmental science, solid waste management, and/or environmental health can supplement the lessons in this book and expose your students to opportunities and knowledge they otherwise would not experience.

The following list suggests people and organizations that may be able to provide more information, visit your class, or arrange for site visits or field trips.

Local businesses and organizations
Check your phone book for contact information for the following:
- College or university civil or environmental engineering and/or environmental science departments
- Hospital infection control nurse or environmental health specialist
- Landfills or incinerators
- Plastics manufacturers
- Recycling companies or programs
- Refuse disposal companies

National and professional organizations
These organizations may be able to provide you with contact information for representatives or members in your area:
- Air and Waste Management Association (412/232-3444; http://www.awma.org)
- American Academy of Environmental Engineers (410/266-3311; http://www.enviro-engrs.org)
- Aseptic Packaging Council (800/277-0888; http://www.aseptic.org)
- Institute of Food Technologists (312/782-8348; http://www.ift.org)
- North American Association for Environmental Education (NAAEE; 937/676-2514; http://www.naaee.org)
- National Association of Environmental Professionals (NAEP; 904/251-9900; http://www.naep.org)
- Polystyrene Packaging Council (800/944-8448; http://www.polystyrene.org)
- Solid Waste Association of North America (SWANA; 301/585-2898; http://www.swana.com)
- State and Territorial Air Pollution Program Administrators/Association of Local Air Pollution Control Officers (STAPPA/ALAPCO; 202/624-7864; http://www.4cleanair.org)

Local government

Check your phone book for contact information for the following:

- County or city engineers
- County or city health department or environmental health division
- Garbage/sanitation engineer
- Sewage treatment department
- Water treatment department

State and federal government

Check your phone book for contact information for the following or visit the World Wide Web sites listed:

- Bureau of Land Management
  - State or local representative or official
  - Environmental education Web site (http://www.blm.gov/education/education.html)
- Bureau of Reclamation
  - State or local representative or official
  - Environmental education Web site (http://www.usbr.gov/env_ed)
- Centers for Disease Control and Prevention (CDC)
  - State or local representative or official; call 404/639-3311
  - CDC Web site (http://www.cdc.gov)
- Environmental Protection Agency (EPA)
  - State or local representative or official
  - Explorers' Club Web site (http://www.epa.gov.kids) Although this club is designed for children ages 5–12, you may be able to adapt the activities and lessons on the site for use with your class.
  - Office of Pollution Prevention and Toxics Web site (http://www.epa.gov/opptintr)
  - Student and Teacher Resources Web site (http://www.epa.gov/epahome/students.htm)

Lesson 1:
Introduction to Solid Waste

In this lesson, students are introduced to the topic of integrated solid waste management through a role-playing scenario. In "**Trash Trouble in Tacktown, Part A,**" students act as citizens of a town faced with a solid waste disposal problem and begin to generate ideas for their roles as specific citizens. As students participate in the lessons in *Understanding Garbage and Our Environment,* they collect ideas and information for carrying out their roles during the town meeting in Part B. (Part B of the scenario is presented in Lesson 8 at the end of this book.)

"**Risky Business**" introduces students to the basic concepts of probability and risk necessary to evaluate the risks and benefits of different waste management methods.

The Garbage Gazette "**Getting a Degree in Garbage,**" included at the end of this lesson, introduces students to the many people who deal with solid waste and the jobs they do.

● Teacher Background on Solid Waste Management

Solid waste management is a very complex issue. Although the three Rs—reduce, reuse, and recycle—are a good place to start, applying them is not always easy. In recent history, humans have been trying to balance the benefits of a healthy environment with the economic costs of achieving those benefits. Different groups have battled over whether costs of certain disposal methods outweigh benefits or vice versa, what solutions should be applied, and who should bear the economic costs. Many factors must be considered when discussing the topic of waste management. Economic, environmental, and personal issues all play major roles in the decision-making process.

In waste management, as with other environmental issues, clear lines do not always exist between what is good for the environment and human health and what is bad. For example, recycling is often promoted as a solution to some environmental problems associated with solid waste disposal. However, companies often run an additional set of trucks to pick up recyclables separately from regular trash. These trucks give off pollutants and use natural resources. Also, in some cases, using recyclable materials instead of lighter and more compact non-recyclable materials may actually compound the problem of limited landfill space, because recycling rates are not 100% for any material in this country.

Because waste management decisions are made within the context of society, personal issues are always a part of the political process. For example, people who live near a proposed landfill or incinerator site are more likely to protest than citizens on the other side of town. This phenomenon is referred to as the "Not in My Backyard" or NIMBY syndrome. Often, disposal sites are located in poorer parts of town where land is cheap and the residents may not have as much political clout as those in more affluent neighborhoods.

The most important idea students should take away from this unit, as the decision-makers of tomorrow, is the realization that often there is no one right answer to complex civil problems. Students should learn that groups of people must learn to work together and to compromise with those who have opposing viewpoints. Waste management plans that work in one region with a certain set of environmental problems, economic factors, and political viewpoints may not be acceptable in another region. As students are introduced to different waste management options in this unit, it is important that they learn to recognize the advantages and disadvantages of each option and that a combination of options may prove to be the best solution to a problem.

● Teacher Notes for "Trash Trouble in Tacktown, Part A"

This activity introduces students to the complexity of waste management decisions through a role-playing scenario. In Part A, students learn the background of the scenario and are assigned roles to play in the Town Meeting. The Town Meeting, Part B of the activity, takes place in Lesson 8 at the end of this unit.

➤ *Due to the nature of this activity, no Student Instructions are provided.*

Group Size .... Class
Time Required .... Getting Ready: 15 minutes
                                                                     Procedure: 30 minutes

Materials

For the Procedure
Per student
• Role Sheet

   Every student should have a different Role Sheet.

Safety and Disposal

No special safety or disposal procedures are required.

Getting Ready

Copy the Roles (provided) and cut them apart. If your class is not large enough to fill all of the roles, you can leave out the ones nearest the end.

Opening Strategy

Explain to the students that although it is important to be aware of the multifaceted issue of solid waste disposal, it is also important to understand that people perceive the issue from different points of view and that every citizen should have an opportunity to present his or her viewpoint in the decision-making process. Tell the students that in order to help them understand these concepts, they will take part in the simulation of a small community dealing with a waste disposal challenge. Remind students of these important points: 1) the focus should be on the process involved in reaching a decision, not on the decision itself, and 2) there is no one right or wrong answer in this simulation or in real life.

Procedure

1. Pass out one copy of Student Information 1 for "Trash Trouble in Tacktown, Part A" (provided) to each student.

2. Give students approximately 10 minutes to read Student Information 1 and to list some possible solutions to Tacktown's problem either in groups or individually.

3. Ask students about some possible ways the people of Tacktown could handle the problem and list student responses on the board. Do not evaluate the responses.

4. Assign roles to the students, passing out the appropriate Role Sheets. (Reserve the role of the mayor for yourself.) The five students assigned the Presentation Roles will give 5-minute presentations at the town meeting in Part B, which is in Lesson 8.

5. Distribute a copy of Student Information 2 (provided) to each student.

6. Inform students that as they perform the investigations over the course of this unit, they should keep in mind the variety of possible solutions for solid waste management in Tacktown. They should gather information, do library research, and interview people to prepare for the town meeting in Part B. Encourage students to seek out others with related roles and work together to form coalitions and come up with solutions. As different activities and demonstrations in this unit are conducted, remind students that they need to be gathering information to support the positions they have been assigned. The five students with Presentation Roles should prepare visual aids summarizing their positions.

Roles
Copy the following roles and cut them apart.

- -

B. Ballott
Teacher Role
Mayor

It is your job to run the town council meeting. You must ensure that everybody who wants to speak has the opportunity and that everyone obeys the rules of common courtesy. As mayor, you are interested in pleasing the majority of the people; after all, elections are coming up. You will probably agree with the majority of the town council. In the case of a tie among town council members, you will consider what the majority of Tacktown citizens want you to do.

- -

I. Might
Town Council Member

Council Member

As a retired accountant, you are most concerned with the financial considerations of any solid waste management decision. You ask questions about money and particularly favor those plans which will cost the taxpayers the least.

- -

M. Bee
Town Council Member

Council Member

You are a retired schoolteacher and now volunteer in Citizens for a Cleaned-Up Environment, a local environmental organization. You are concerned with air and water quality. You ask questions about these issues and will not vote for a plan unless it seems to be environmentally sound.

- -

Council Member

P.R. Haps
Town Council Member

You are the owner of a small business in town and a member of the Chamber of Commerce. You are particularly interested in the rights of small businesses. You ask how the different plans may affect these small businesses and prefer to vote for plans that seem to be in their best interest.

Council Member

I. Hedge
Town Council Member

You have a background of public service. You are concerned about how the voters feel on this issue and are particularly concerned about unemployment. Your vote will favor plans that seem to please the most people and put the fewest people out of work.

A. Abacus
Town Treasurer

As town treasurer, it is your responsibility to keep track of the costs of the different plans. As the different projects are presented during the meeting, write the total costs of waste pickup on the chalkboard. Be sure to write down the total cost of each plan, especially when more than one company is dealing with separate parts of the waste stream.

D. Digger
Presentation Role

Landfill Manager

You have owned and managed Little Landfill for the past 40 years. You are well known and well liked by most community members. Presently, your landfill accepts waste only from residents and businesses in Tacktown and the surrounding township. You estimate that at the present fill rate, Little Landfill will be closed in four years—five years if the 25% reduction in the solid waste stream actually takes place a year from now. You have selected and purchased land for a new landfill site to use when Little Landfill closes. Your landfill company employs 25 people and can continue to do so as long as the cost of waste hauling remains the same after the 25% reduction goes into effect. If the community insists on lower solid waste disposal costs, you will have to lay off five people. Your current charges for waste pickup are $6 per month for town residents, $10 for township residents, and $40 for small businesses with dumpsters. You haul waste away from large businesses and construction companies at a cost of $200 per month.

A. Genn
Presentation Role

Recycling Company Representative

You have run Repeat Recycling for the past five years in Next Door City and are looking to expand. You are willing to implement curbside recycling in Tacktown at a monthly cost of $2.50 per town resident, $3.50 for township residents, $20 for small businesses, and $100 for large businesses. To accommodate this expansion, you would need to hire three new workers. For no fee, you would be willing to leave several drop-off box trailers in Tacktown in which citizens could drop off their own presorted recyclables.

N. Mix   Presentation Role
Material Recovery Facility Representative

You are considering building a Material Recovery Facility (MRF) in Medium County. In order for it to be economically feasible, you need the cooperation of all three communities. You already have some interest from Neighborville and Next Door City but really need the commitment of Tacktown to get financial backing. You want to take all of the solid waste from Tacktown and run it through your MRF facility to remove the non-contaminated recyclables. The remainder of the waste would go to Little Landfill for disposal. In order to successfully operate your MRF you would need to charge $8 per town resident, $13 for township residents, $60 for small business dumpsters, and $300 per month for large businesses. Out of this money, you would pay Little Landfill 20% of what you make to take the waste after you have separated out the recyclables. You estimate that 30% of the materials in the waste stream will be removed in your facility. You would hire five new people to work in your facility.

M. Mulcher   Presentation Role
Composting Company Representative

You have owned and managed Column Composting in Neighborville for the past three years. It is a small operation that produces enough mulch to provide for the community's needs as well as some outside homeowners, landscapers, and nurseries. You would be willing to accept and transport any of Tacktown's compostable waste free of charge, as long as it was already sorted out from the waste stream, free from contaminants, and gathered in a central location. If you were to receive all of Tacktown's compostable waste, you would hire one new person.

B. Burns   Presentation Role

Incineration Company Representative

You would like to build an incinerator in Tacktown to accept their solid waste as well as that of Neighborville and Next Door City. You have financial backing for your plan; all you need is a market. Neighborville and Next Door City have both agreed to send you all of their solid waste so long as the incinerator itself is located in Tacktown. You need approval from the town in order to get a building permit. You would accept waste from the town for the monthly price of $8 per town resident, $12 per township resident, $55 for small business dumpsters, and $275 for large businesses. You would install scrubbers and electrostatic precipitators to ensure air quality. Out of this money you would pay Little Landfill 10% to take the ash left over after the incineration process. Because incineration reduces the volume of solid waste by 90%, this would increase the life of the landfill by up to 30 years. You would employ 10 new people.

Steve S.

P. Pulp

Paper Company Representative

You are a representative of Pretty Paper, Inc., which has been one of Tacktown's largest industries for the last 40 years. You would be willing to accept paper from the town for recycling if it was sorted from the waste stream into different paper types and was free from contaminants. Although you would incur some initial cost for installing the equipment to handle the recycled paper pulp, you would be able to absorb that cost and want to show the community your support. You also hope that being able to display the recycled paper symbol on your products will boost your sales.

B. Bond
Chemical Company Representative

You have been the owner and operator of Bond Chemical Corporation in Tacktown for the last 20 years. Your hazardous wastes do not go to Little Landfill because it is not certified to take them, but you do send office and cafeteria waste there. Currently you must transport your hazardous waste to Hazardous Hole in a neighboring county. If a new landfill is built in Tacktown, you hope the appropriate permits for a secure landfill will be obtained and you can send all of your waste there. This would save you huge transportation costs. You are trying to persuade D. Digger to make the new landfill a secure one. D. Digger has informed you that it would not be economically feasible to put nonhazardous waste into a secure landfill because of the high costs of the permits and burial methods. Therefore, you would need to find enough hazardous waste business to make it worthwhile. You believe you might succeed and therefore would like the new landfill built no matter what is decided about the other solid waste issues.

G. Grows
Farmer

You and your relatives have been farming the land neighboring Tacktown for decades. Because you live outside the city limits, trash pickup for you costs almost twice what it does for the town residents. You are concerned that the new solid waste disposal plan will increase your costs even more. You currently send your organic materials to Column Composting and use their compost on your garden. You would like a plan that will not increase your pickup costs or require you to drive all the way into town to drop items off.

P. Proffitt
New Business Interest

You represent a large business that is considering Tacktown as a site for one of its new plants. Building your business here would bring a lot of money into the community. You are not willing to sign a contract of sale until a solid waste management program has been decided upon that is fair to businesses because you do not want to pay an unfair portion of the community's new solid waste bill. You consider anything over $250 per month in combined solid waste removal bills to be unfair and it would dissuade you from building a plant in Tacktown. You would hire 50 employees from the three communities in Medium County.

L. Lite
Power Company Representative

You represent the Litebrite Electric Company, which supplies all of the power to Tacktown, Neighborville, and Next Door City. You would like to see the incinerator built so you would have an additional energy source so that the increasing energy demands of the community can be met. You have already worked out a deal with the incinerator company to buy the energy, and you are trying to rally community support.

M. Whett
Water Company Representative

As the manager of the Puregood Water Purification Plant, your main concern is the demand on Tacktown's water supply. The paper company already uses a large portion of the available water, and you are afraid that if they start using recycled paper to make new paper, they will need even more water. You are especially concerned about increased use of water to remove impurities from the paper before recycling. An additional concern is the cost to remove these impurities from the water before it is returned to Tacktown's water supply.

S. Small
Chamber of Commerce Small Business Representative

As the Tacktown Chamber of Commerce representative, you are particularly concerned about the small businesses in Tacktown. You feel that the new solid waste management plan might be too hard on some of the financially unstable companies and might put them out of business. They already pay $40 per month in dumpster fees. You emphasize that it is the small businesses that are the tradition of Tacktown and the heart of the community. They should not be expected to pay any more. You feel the big businesses, like the paper mill, should bear more of the cost for the new solid waste management plan because they contribute more to the total waste volume.

G. Green
Environmental Group Representative

Your main concern in this issue is maintaining the benefits of a healthy environment; money should not be as much of a concern when deciding on a solution. You do not support a Material Recovery Facility because you believe too much contamination of recyclables takes place, making them unusable. You do not support an incinerator because you are concerned about declining air quality. You believe the best solution is probably a combination of composting and curbside recycling.

N. Nozie
Landfill Neighbor

You have lived downwind from Little Landfill all of your life. You are disturbed by the odor, the traffic, and the litter that sometimes blows out of the trucks. However, although D. Digger has offered to buy your house from you at twice its market value, you refuse to leave under any circumstances. You will support any plan that reduces the amount of garbage truck traffic to Little Landfill.

N. Nimby
Proposed Landfill Site Neighbor

You live right next to a parcel of land D. Digger has purchased as a site for the next landfill after Little Landfill closes. You plan to retire and move to Florida in 10 years and do not want the landfill there within that time period. You are also concerned that you will not be able to sell your home because it is so close to a future landfill site. You support whichever plan would extend the life of the current landfill the longest.

C. Choke
Citizen Concerned about Air Quality

As a community member with asthma, you are particularly concerned about air quality. You have heard that incinerators release toxic chemicals and are concerned about your health and the health of other community members. You will not support a plan that includes incineration until you are convinced that it will not have a negative impact on your health.

I. Gulp
Citizen Concerned About Water Quality

As a community member who enjoys fishing in the streams just outside of Tacktown, you are particularly concerned about water quality. You have heard a rumor that Bond Chemical Company has been sending hazardous chemicals to the landfill, and you are concerned that these chemicals have been leaking into area rivers and streams. You feel this matter of surface water contamination should be resolved before any decision on a solid waste management plan can be made.

D. Gooder
Citizen Supportive of Recycling

You have been recycling in your own home for years and think everybody should do the same. You believe this solid waste reduction plan is the perfect opportunity to educate community members about the benefits of recycling. You do not believe MRF recycling would be a good idea because of the resulting contamination of resources and because it encourages community members to leave their recyclables in with their trash and let somebody else deal with the problem. You feel the large companies in the community should shoulder most of the cost for implementing a new recycling program and educating the citizens of the community about the new program.

I. Dell
Unmotivated Citizen

You feel it is your right as a tax-paying citizen in the community to put your trash out at the curb and have somebody else deal with the problem. You want the community to adopt a plan that does not involve any additional effort on your part. You do not want to separate your trash or have to drop things off anywhere. You do not mind if, as a last resort, your taxes have to be raised, but you feel the businesses in the community should bear most of the cost.

A. Roma
Compost Company Neighbor

You live close to the Column Composting plant in Neighborville. You have complained about the smell of the plant and hope to have it closed down. Although officials have assured you that no unhealthy amounts of pollutants are present in the air and your health is not at risk, you are convinced that living near the plant may harm your children. You are here to rally support against the compost plant and to convince the citizens of Tacktown not to send any of their material there.

N. M. Ploid
Citizen out of Work

The small factory you worked for recently laid you off. You and several of your former co-workers are receiving unemployment benefits and are looking for work. You have not had much luck in the job market and are concerned about how much longer your unemployment benefits will last. You do not want to see your family go hungry. You do not understand how people could put environmental issues before employment. You support any program that you feel gives you and your friends the best chance of finding work.

I. Haul
Truck Driver

You and your family have just moved to Tacktown to be closer to your elderly parents. You have had many years of experience as a truck driver and are looking for a job. You would like to avoid taking a position with your old firm because it involves driving across several states and spending many nights on the road. If you can get a job driving locally, you can stay at home and spend more time with your family. You would support any program that would increase the need for truck drivers in the area.

I. Bury
Representative from Hazardous Hole

You have managed Hazardous Hole Secure Waste Facility for the past 20 years. During this entire time, Bond Chemical Corporation has been one of your biggest customers. You heard that B. Bond is at the town meeting pushing for a local secure landfill to help defray transportation costs. If you lost Bond Chemical Corporation's business, you are afraid that you may go out of business, leaving all of your workers unemployed and leaving your other customers with no means of disposing of their hazardous waste. In order to keep B. Bond's business, you would be willing to negotiate transportation fees.

L. Law
Police Officer

You have been a police officer in Tacktown for 5 years. You feel one of the most frustrating things about your job is enforcing dumping laws. There are several members of the community who litter while traveling in their cars, or even worse, dump their household refuse over the bluffs at the edge of town. You are concerned that if disposal costs become greater, you and your fellow officers will have an even more difficult time preventing this littering. You are therefore interested in supporting programs which will keep total disposal costs low.

I. Cure
Hospital Administrator

You are the administrator for Tacktown Hospital. Part of your job is finding the most economical means for disposing of the hospital's biohazardous waste. Currently, this waste is transported some distance to be incinerated at a site in a neighboring county. You support the construction of a local incinerator to reduce your transportation costs. You are concerned about the risk of a biohazardous exposure during the long distance the items are currently being transferred and feel reduced risks would be another benefit of local incineration.

G.O. Vern
Member of the EPA

You are a member of the Federal Environmental Protection Agency. You came to the town meeting to provide accurate information about environmental regulations. You want to be sure that whatever waste disposal program the community adopts is in compliance with regulations about landfill construction, landfill monitoring, and the appropriate disposal of toxic chemicals.

X. Periment

Scientist

You are a chemistry professor at a nearby university. You know that often people have difficulty interpreting scientific data. You have come to the town meeting to offer your services and share your expertise in risk assessment. You are particularly concerned that the town members may not have an understanding of the types of risks they take every day, such as driving an automobile. You can provide them with some basic information about risks. You feel once they have this information, they will be better prepared to compare the risk assessment information that may be presented during the meeting.

● Student Information 1 for "Trash Trouble in Tacktown, Part A"

Description of Tacktown

Imagine that you are a citizen of an imaginary community called Tacktown. Tacktown is a community of 25,000 people. It is located in Medium County along with two other towns: Neighborville (population 20,000), and Next Door City (population 30,000). The rest of Medium County consists of farms and a state park.

Recently, solid waste disposal has become a hot issue in Tacktown. Big State has mandated that Tacktown and all other communities in the state reduce the solid waste stream entering their landfills by 25% within three years. Currently, all of Tacktown's municipal solid waste goes to Little Landfill for disposal.

You have just found out that a town meeting has been scheduled to discuss Tacktown's solid-waste management strategy. Several waste management industries, local businesses, and community members will present their opinions on Tacktown's dilemma at the town meeting. The community must decide how to meet the state's mandate through integrated waste management, which balances different options and doesn't rely solely on one disposal method.

Tacktown has a variety of businesses, such as Pretty Paper, Inc., a large paper company; Bond Chemical Corporation; and a variety of small businesses. The electricity for the community is supplied by the Litebrite Electric Company, and the water goes through the Puregood Water Purification Plant. Repeat Recycling is located in Next Door City, and Column Composting is located in Neighborville. Hazardous Hole, a hazardous waste disposal facility, is located in a neighboring county.

● Student Information 2 for "Trash Trouble in Tacktown, Part A"

Engineering and Economic Report

The overall solid waste stream breakdown for Tacktown and the surrounding township is shown in the following graph:

Sources of Waste in Tacktown

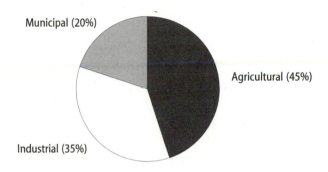

The following table shows how this waste is currently disposed of:

Tacktown Solid Waste Disposal				
	Destination of Waste			
	Column Composting	Little Landfill	Repeat Recycling	Hazardous Hole
Agricultural	90%	10%	0%	0%
Industrial	0%	90%	5%	5%
Municipal	5%	90%	5%	0%

The Tacktown engineering team conducted an inventory of just those wastes being sent to Little Landfill for disposal. Their results are shown in the following table:

Little Landfill Waste Inventory	
Type of Waste	Percentage by Weight
paper and paperboard	37.1%
yard wastes	19.9%
rubber, leather, textiles, wood	10.1%
metals	9.7%
plastics	8.3%
food wastes	5.9%
glass	5.4%
miscellaneous inorganic wastes	3.6%

The latest Tacktown census reported the following employment rates:

Tacktown Employment Rates	
Business	Number of Employees
Small Businesses (combined)	2,192
Pretty Paper, Inc.	1,513
Bond Chemical Corporation	1,255
Service Industries	754
Puregood Water Purification Plant	35
Litebrite Electric Company	30
Little Landfill	25

● Teacher Notes for "Risky Business"

This activity demonstrates the basic concept of risk. Students compare risks and benefits and make decisions based on their evaluation.

> *Part A covers the basic concept of probability, while Part B is a little more advanced and asks students to make decisions based on their understanding of probability. Part B may be frustrating for younger students.*

Group Size .... 3–4 students
Time Required .... Getting Ready: 10 minutes
　　　　　　　　　　　　　　　　　　　　　　　Procedure: 30 minutes

Materials

For Getting Ready
Part A, per group
- opaque plastic cup
- label or piece of masking tape
- marker or pen
- 8 red beads
- 6 blue beads
- 6 yellow beads

Part B, per group
- 6 film canisters
- 6 labels or pieces of masking tape
- 9 red beads (in addition to the ones from Part A)
- 9 blue beads (in addition to the ones from Part A)
- 8 yellow beads (in addition to the ones from Part A)
- 7 orange beads
- 6 green beads
- 5 purple beads
- 8 gray beads
- 8 black beads

For the Opening Strategy
Per class
- large coin, such as a quarter
- 6-sided die (from a pair of dice)

For the Procedure

Part A, per group

• Community Resources Cup prepared in Getting Ready

Part B, per group

• materials from Part A
• 6 Resource Canisters (prepared in Getting Ready)

Getting Ready

For Part A

1. Label each opaque cup "Community Resources Cup."

2. Put 8 red beads, 6 blue beads, and 6 yellow beads in each Community Resources Cup.

For Part B

1. For each group, label and fill six film canisters with beads as described in the following table:

Table 1: Resource Canisters	
Canister Label	Contents
Food: 10% chance of illness	9 red, 1 gray
Water: 10% chance of death	9 blue, 1 black
Shelter: 10% chance of illness, 10% chance of death	8 yellow, 1 gray, 1 black
Comfort: 20% chance of illness, 10% chance of death	7 orange, 2 gray, 1 black
Wealth: 20% chance of illness, 20% chance of death	6 green, 2 gray, 2 black
Political Power: 20% chance of illness, 30% chance of death	5 purple, 2 gray, 3 black

Opening Strategy

For Part A: How Good Are Your Chances?

1. Ask the class if they know what "probability" is. Younger students may be more comfortable using the word "chance." Define probability (or chance) as a number expressing how likely it is that a specific event will occur. Probability is calculated by dividing the number of chances of a particular event occurring by the total number of chances available.

2. Flip a coin for the students. Explain that there is one chance that the coin toss will turn up heads out of the two total number of chances. The coin will turn up heads or it will turn up tails: $1 \div 2 = 0.5 \times 100\% = 50\%$ chance of the coin turning up heads. Thus there is a 50% probability that it will turn up heads and a 50% probability that it will turn up tails.

3. Now roll the die. The die has six numbered faces, so the probability of any one number coming up is $(1 \div 6) \times 100 = 0.167 \times 100 = 16.7\%$ that a certain number will come up. If you wish, keep the probability in terms of fractions (such as ½ instead of 50%) for students who are less familiar with percentages.

   Explain to students that sometimes different options can be grouped together and the probability can be calculated for the entire group. For example, the probability of a roll of the die being 4 or less is $(4 \div 6) \times 100 = 0.67 \times 100 = 67\%$ (⅔); whereas, the probability of it being 5 or 6 is $(2 \div 6) \times 100 = 0.33 \times 100 = 33\%$ (⅓).

4. Give each group a Community Resources Cup and each student the Student Instructions. Have them do Part A of the Procedure in the Student Instructions.

For Part B: Is It Worth the Risk?

1. Be sure students understand the difference between the probability of selecting a bead of a certain color and the actual selection of beads in Part A of the Procedure. Have students consider the probability of selecting a yellow bead from the cup. There are 6 yellow beads out of 20 total beads, so the probability of selecting one is $(6 \div 20) \times 100\% = 30\%$. Out of every 10 beads sampled, 3 should be yellow. However, looking at the class data it is highly unlikely that every student had 3 yellow beads in the 10 they selected. Probability refers to likelihood or chance of an occurrence. Small samples may have significantly different ratios from those probabilities. However, the bigger the sample, the closer the ratio will be to the calculated probability. For example, while each student's ratios of selected colors may or may not be close to the estimated probabilities, each group's average should be closer to those probabilities, and the entire class average should be closer still.

2. Ask students if they know what risk is. *The danger or probability of suffering loss, damage, or hazard.* Risk is like probability, but in this case, at least one possible result is something that is considered harmful. Because "harmful" is a value judgment, calculating whether or not a risk is worth taking can also be a value judgment. To illustrate this concept, pose the following scenario: Suppose scientists developed a new medicine that cured a very serious and debilitating disease. This new medicine was tested and cured 99% of those people tested;

however, 1% did not survive the treatment. If somebody was suffering from this illness, he or she would need to weigh the potential benefit of being cured against the potential risk of not surviving. Because the risk of not surviving is only 1%, many people may choose to try the cure.

On the other hand, suppose a new drug was developed that could improve a person's intelligence significantly. Unfortunately, 50% of the time it also left the person paralyzed. Most people would be reluctant to try it because the benefit of being smarter would have to be weighed against the very high risk, 50%, that they would end up paralyzed.

3. Have students read the Student Background (provided). As a class, discuss why some risks may be more acceptable to some people than others.

4. Give each group six Resource Canisters and have them do Part B of the Procedure.

Discussion

1. Determine the game winner(s). Have them explain what Resource Canisters they chose and why.

2. Discuss with the students what value judgments they used to evaluate risk. Was the risk associated with basic needs such as food, water, and shelter more worth taking than risks associated with the extras such as comfort, wealth, and political power? What might some of these risks be in real life? (For example, what risks might be associated with food, shelter, or comfort?) Have the students answer questions c and d in the Analysis Questions section.

Extension

For older students, this lesson can be taken to a more mathematically and statistically advanced level. Have students predict the outcomes if a smaller or larger sample (fewer or more beads per student) were taken. How would this change the mathematical probability of each community receiving the basic resources they need to survive? Would they be willing to take more or less risk as a result?

Cross-Curricular Integration

Social studies:
- Introduce a discussion of opinion polls, random samples, and margins of error in public policy or a decision-making process.

Explanation

New medicines, products, and inventions all need to be evaluated for public safety before they are allowed on the market. Very few things can be considered 100% risk-free, so the potential benefits of a new product must be compared to the potential risk.

Decisions affecting the environment undergo the same process. The potential benefit of a proposed method of waste disposal, such as incineration or landfilling, must be evaluated using risk-benefit analysis. In the case of incineration, the benefits of reduced volumes of waste and the potential for harnessing energy need to be weighed against the potential adverse human health effects from toxic air emissions or incinerator ash.

● Student Background for "Risky Business"

Different methods of waste disposal have different risks associated with them. For example, the incineration process can potentially cause air pollution. Landfills can potentially cause water pollution if harmful chemicals leak out. Potential health risks associated with waste disposal methods must be considered when choosing between these methods. In order to evaluate these risks fairly, it can be helpful to compare them to other, more familiar risks. The tables on the next page list some of the everyday risks Americans face. Some risks are voluntary (taken by choice), such as those associated with smoking or driving a car; other risks are involuntary, such as background radiation or being struck by lightning. The Risks of Death from Air Pollution table shows the average lifetime risk due to air pollution from polycyclic organics (a rough indicator of incineration emissions), and specific risks attributed to the incineration of hazardous wastes.

A risk from a particular source of exposure is usually quantified as a numerical expression that looks like this (remember that $10^3 = 10 \times 10 \times 10 = 1,000$ and that $10^{-3} = 1/1000$):

$$\textit{lifetime risk of death from a home accident} = 7.7 \times 10^{-3}$$

$$\textit{where } 7.7 \times 10^{-3} = 7.7 \times \frac{1}{10^3} = 7.7 \times \frac{1}{1000} = \frac{7.7}{1000} = 7.7 \textit{ deaths per thousand people}$$

This means that an average of almost eight (7.7) out of every 1,000 Americans will die in a home accident. Similarly, a risk of 2.2×10^{-2} means that an average of 2.2 out of every 100 people will die from a given cause (remember that $10^2 = 10 \times 10 = 100$).

Carcinogens (cancer-causing substances) with an individual risk level of 4×10^{-3} or above (four chances or more in 1,000 that a chronically exposed individual may develop cancer) are regulated. Chemicals with individual risk levels as low as 1×10^{-6} (one chance in 1 million that a chronically exposed individual may develop cancer) are not usually regulated. The risks discussed here refer to lifetime risks, although sometimes risks are represented as annual risks.

Analysis of risk is an important part of public decision-making. Communities often debate the dangers associated with environmental issues when some risk of adverse health effects to humans is involved. For example, risk analyses help determine whether incineration plants should be built and, if so, where they should be built. Risk analysis would also help scientists and engineers determine what types of pollution control devices were necessary and what kinds of waste would be accepted at the incinerators.

Risks of Death from Accidents

Cause of Death	Lifetime Average Risk	Deaths per 1 Million
motor vehicle accidents	1.7×10^{-2}	17,000
all home accidents	7.7×10^{-3}	7,700
firearms	7.0×10^{-4}	700
electrocution	3.7×10^{-4}	370
lightning strike	3.5×10^{-5}	35

Risks of Death from Cancer

	Cause of Death	Lifetime Average Risk	Deaths per 1 Million
	Cancer, all types	1.9×10^{-1}	190,000
Radiation	being an airline pilot	2.8×10^{-3}	2,800
	medical x-rays	1.4×10^{-3}	1,400
	background radiation	1.4×10^{-3}	1,400
	flying 4 hours per week	7.0×10^{-4}	700
	living in a masonry building	3.5×10^{-4}	350
	flying once per year	7.0×10^{-5}	70
Eating and Drinking	alcohol, 1 beer per day	1.4×10^{-3}	1,400
	one 12.5-ounce diet drink per day	7.0×10^{-4}	700
	saccharin (average consumption)	1.4×10^{-4}	140
	half a charcoal-broiled steak per week	2.1×10^{-5}	21
Tobacco	smoker, all effects	2.1×10^{-1}	210,000
	smoker, cancer only	8.4×10^{-2}	84,000
	person living with smoker	7.0×10^{-3}	7,000

Risks of Death from Air Pollution

	Cause of Death	Lifetime Average Risk	Deaths per 1 Million
	Air pollution (polycyclic organics), all effects	1.1×10^{-3}	1,100
Hazardous Waste Incinerator Emissions	dioxin and furan emissions	1.2×10^{-5}	12
	PIC[1] emissions	2.7×10^{-6}	2.7
	principle organic hazardous constituents (POHCs)	1.0×10^{-6}	1.0
	polychlorinated biphenyls (PCBs)	5.2×10^{-8}	0.052

[1] When hazardous organic wastes are incinerated, new organic compounds can form outside the heating zone due to the recombination of molecular fragments. These products of incomplete combustion have been termed PICs.

● Student Instructions for "Risky Business"

This activity demonstrates the basic concept of risk. You will compare benefits to risks and make decisions based on your comparison.

Procedure

Part A: How Good Are Your Chances?

1. Carefully pour the beads onto a table or desk. Count the number of each color of bead and record it in the appropriate space in the Color Probability table in the Data Recording section of this handout. Count the total number of beads and record that number in all three rows of the appropriate column.

2. Calculate the probability of randomly selecting each color of bead from the mixture of beads, following the example on the Color Probability table for each color. After you have calculated the probability of selecting each of the three colors, return all of the beads to the cup. (The sum of the probabilities for the three colors should be 100%.)

3. Taking turns, have each group member select one bead from the cup without looking. The first student to select a bead is student A. Student A then records the color of the bead he or she selected by making a mark in the appropriate color row under column A on the Color Trial Results table in the Data Recording section. He or she then returns the bead to the cup and shakes it up before the next student takes a bead. Then students B, C, and D each select one bead in turn, also recording the color, returning the bead to the cup, and shaking it.

4. Continue taking turns selecting, recording, and returning beads until everyone has had 10 turns. Each student should calculate the percent of each bead color he or she selected and record these values on the chart. Then calculate the group's average for each color and record those values.

5. Answer question a in the Analysis Questions section of this handout.

Part B: Is It Worth the Risk?

1. Have one member of your group read aloud the first two columns of the Community Resources Probability table in the Data Recording section to tell the group members what each color of bead represents and what its point value is. (Notice that illness and death have negative values.)

2. Look at the beads in each Resource Cup. These beads represent both the risks and benefits associated with each resource. Notice that in addition to its resource beads, each cup contains at least one gray or black bead—some contain both, and some contain more than one of both. Remember that gray beads represent illness and black beads represent death.

3. Imagine that your group actually represents a city-sized community that must work together to accumulate resources and compete with other communities. You can add to your Community Resources Cup any or all of the six Resource Canisters, but you must add the entire contents of each chosen canister—including gray and black beads. Consider the following rules:
   * If you do not draw at least two red, two blue, and two yellow beads during the game, someone in your community lacks these basics of survival, and that community member is dead. Each occurrence of death receives the same point value (–20) as a black bead.
   * If you selected one or more gray beads during the game, you must also have selected at least three red, three blue, and two yellow beads for each gray bead you selected, or the sick member(s) of your community will not survive the illness without the extra basics of life. Each occurrence of death receives the same point value (–20) as a black bead. (This point value replaces the –10 for illness, so don't deduct both death and illness.)
   * If you selected one or more black beads during the game, one member of your community has "died" for each black bead.
   * The community with the most points at the end of the game wins.

4. As a group, choose which combination of resources you want for your imaginary community, and add the entire contents of these canisters to your Community Resources Cup, along with the beads already in the cup from Part A. Answer question b in the Analysis Questions.

5. Pour all the beads carefully onto a table or desk and count the total number of beads of each color. Use this information to fill out the Community Resources Probability table. Follow the example and calculate the probability of selecting each color of bead. Return the beads to the cup.

6. As in Part A, take 10 turns each selecting one bead from the cup, marking it in the appropriate column of the Community Resources Trial Results table, and returning the bead to the cup. Add up your individual data in the column with your letter (A, B, C, or D), taking into account the penalties mentioned in the rules (see step 3) for not having enough food, water, and shelter. (Point values are listed in the Community Resources Probability table.) As a group, add your point values to determine your community's score.

Data Recording

Table: Color Probability

Bead color	Number of beads this color	Total number of beads	Probability of selecting this color of bead (in percent)
Example: white	5	40	(5 ÷ 40) x 100 = 0.125 x 100 = 12.5 %
red			
blue			
yellow			

Table: Color Trial Results

Bead color	Number of beads this color selected per student				Total number of beads selected per student	Percent of beads selected that color represents				
Student	A	B	C	D		A	B	C	D	average
Example: white	2	4	1	3	10	20%	40%	10%	30%	25%
red										
blue										
yellow										

Table: Community Resources Probability

Bead Color	Point Value	Number of beads this color	Total number of beads	Probability of selecting that color of bead in percent
Example: white		16	60	(16 ÷ 60) x 100 = 0.267 x 100 = 26.7%
red: food	10			
blue: water	10			
yellow: shelter	10			
orange: comfort	15			
green: wealth	15			
purple: political power	20			
gray: illness	-10			
black: death	-20			

Table: Community Resources Trial Results

Bead color	Number of beads each color selected per student			
Student	A	B	C	D
Example: white	2	4	6	4
red: food				
blue: water				
yellow: shelter				
orange: comfort				
green: wealth				
purple: political power				
gray: illness				
black: death				
Total Points				

Analysis Questions

a. How do the individual and average percentages of colors picked compare to the probabilities of selecting each color calculated in step 2 of the Procedure? What could be done to make these values even closer?

_____

_____

_____

_____

b. Which resources did you add to your Community Resource Cup?

_____

_____

_____

c. In Part B, how did your community do in terms of survival? Did anyone "die" or "get sick"?

_____

_____

_____

d. If you could do Part B again, how could you change your strategy to reduce the risks of illness and death?

_____

_____

_____

_____

The Garbage Gazette

October 9   Local Edition   Vol. 1, Issue 1

Getting a Degree in Garbage

The cliché "It's a dirty job but someone has to do it" is never more appropriate than when that job relates to garbage. From the time refuse is picked up from the curb until it arrives at its final resting place, a variety of people in a multitude of jobs have had a hand in what happens to your trash. Who are these people and what kinds of jobs do they do?

Perhaps the most visible career in waste management is that of "garbage collector." Whether you have curbside service or a dumpster, the garbage collectors collect your garbage and deliver it to the landfill, recycling center, waste-to-energy facility, or whatever disposal facility your city uses.

At these disposal facilities are sorters, graders, engineers, scientists, and others who make sure the disposal process runs smoothly and efficiently. They look over the waste to check for any materials that should not be disposed of in their facility. These materials are then removed and disposed of properly. They also monitor leachate and other emissions to ensure that these are kept within EPA standards.

The facilities didn't just appear overnight, of course. So who decides what disposal facility is best for your town? Who decides where to put it? Who designs it?

The city government works with either an elected or appointed official like a city planner or with a con-

Trained professionals are responsible for making sure your garbage is disposed of properly.

sultant to find the type of facility that best fits the needs of the town. The official or consultant usually has a college degree in environmental engineering and/or years of experience in the field. Architects and engineers who specialize in waste disposal facilities are hired to design the actual facility itself using the recommendations from the official or consultant, the city government, and the citizens themselves.

Because proper waste disposal is important to our health and the health of our environment, people who work in the waste disposal in-

dustry are specially trained to do their jobs. Apprenticeships and on-the-job training give people hands-on experience. Conferences and workshops sponsored by organizations such as SWANA (Solid Waste Association of North America) or the Academy of Environmental Engineers gives industry people a chance to talk to others in the field and learn about cutting-edge techniques being used in other parts of the country. Colleges and universities offer classes and degrees in waste management, environmental engineering, and environmental chemistry to prepare people for jobs in the industry.

So the next time you take out the trash, don't think of it as a dirty job—think of it as a chance to work with the pros.

Think About It

1. What are some different jobs that people have in the solid waste industry ?

2. Who decided what kind of waste disposal facility your town has? If this person is an elected or appointed city official, what other kinds of decisions about the city does he or she make?

3. If you wanted to learn more about the solid waste industry and the careers involved, who in your city could you contact to answer your questions?

Lesson 2:
Waste Characterization

Students investigate the concept of waste characterization through "**What's Waste?**," a hands-on inventory of classroom waste aimed at getting students to think about what they throw away. The results of the solid waste inventory are compared to nationwide averages from information sheets (provided) that can be used as overheads or student handouts.

No matter how waste is characterized, it has to be disposed of properly. *The Garbage Gazette* "**Garbage Takes a Vacation**" retells the story of the *Mobro 4000,* the notorious garbage barge that couldn't find a place to dock in 1987. In a second *Garbage Gazette*, "**Archaeologists Find Garbage Tells the Truth**," students are introduced to the people who study our garbage after it's thrown away and the lessons it can teach us about our lifestyles.

● Teacher Background on Waste Characterization

Everyone produces trash; garbage collected at home, in school dumpsters, and in roll-off boxes at construction sites must be disposed of. Humans have always produced trash and have always disposed of it in some way, so solid waste management is not a new issue. What has changed are the types and amounts of waste produced, methods of disposal, and human values and perceptions of what should be done with it.

Most people do not spend much time thinking about what types of materials they throw away or wondering about the exact makeup of the contents of a garbage truck. But if you were to ask what category of material they thought the biggest portion of the contents fell into, you would get many different responses. Perceptions of the makeup, or characterization, of the solid waste stream are affected by many factors, including personal consumption, media reports, and visual impressions of litter and overflowing trash cans. However, we do not need to rely on these often inaccurate sources to learn about the solid waste stream. The EPA and other government agencies compile data and report on the actual contents of our national municipal solid waste (MSW) stream, so we can compare our perceptions with real data. Key information from the EPA's 1995 report is summarized in the following graph, which shows data for both materials and products.

The Contents of Municipal Solid Waste,* by Product and Material

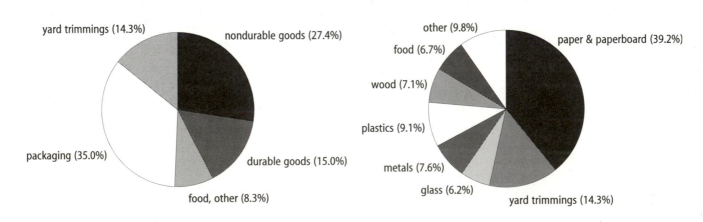

yard trimmings (14.3%)
nondurable goods (27.4%)
packaging (35.0%)
durable goods (15.0%)
food, other (8.3%)

other (9.8%)
food (6.7%)
wood (7.1%)
plastics (9.1%)
metals (7.6%)
glass (6.2%)
paper & paperboard (39.2%)
yard trimmings (14.3%)

Products

Materials

*208.0 million tons in 1995

The MSW characterized in the EPA report includes waste from residential, commercial, institutional, and industrial sources. ("Industrial" waste in this category is packaging and administrative waste, not process waste.) Other kinds of waste, such as agricultural waste and municipal sludge, are not addressed in the report.

The terms "solid waste," "refuse," "garbage," and "trash" are often used interchangeably. However, solid-waste professionals distinguish between them. Solid waste and refuse are synonyms and refer to any of a variety of materials that are rejected or discarded as useless. These terms are subjective, since they rely upon whether or not humans perceive the objects to have any use or value. The variety of materials referred to as solid waste or refuse is broken into two subcategories. Garbage refers strictly to animal or vegetable wastes, particularly by-products of food preparation. It creates offensive odors and decomposes rapidly if exposed to the elements. Trash refers to everything else. Since most of us do not make distinctions between the terms "garbage" and "trash" in our everyday language, they will be used interchangeably throughout this book.

Most communities now deal with their solid waste using an integrated waste management system. Such a system combines two or more of the following processes:

- Source reduction—Source reduction keeps materials from entering the waste stream. Less waste is created; therefore less waste needs to be disposed of.

- Reuse—Reusing materials postpones their entry into the waste stream.

- Recycling—Recycling decreases the amount of material that must be landfilled. It is classified into two major approaches: source separation, in which each household or business separates the recyclable materials by type so they can be collected separately, and materials recovery facility (MRF) recycling, in which all recyclable materials are sent together to a facility where they are separated by manual and technological means.

- Composting—Composting, the controlled biological degradation of organic materials, recovers certain materials, mainly garbage, from the waste stream and generates a valuable product. The organic material biodegrades when oxygen, moisture, and microbes are present.

- Incineration—Incineration is the controlled high-temperature burning of combustible materials and can recover some of the energy in solid waste. Incinerators generate power from the resulting heat and can lead to a 90% reduction in the volume of material to be landfilled.

- Landfilling—Landfilling is the practice of burying waste. Landfills differ from dumps in that each day's waste is covered with soil, and various technologies are used to prevent the spread of contaminants and the buildup of explosive gases. Landfills will always be needed to dispose of wastes that cannot be safely or economically handled by other means.

Integrated waste management considers different waste management needs and fits available waste management options to a city's or region's needs. The Environmental Protection Agency estimates that of the total solid waste stream in the United States, 56.9% is landfilled, 27% is recycled, and 16.1% is incinerated. Because waste management is the third highest cost to local governments, each waste item should ideally be matched to the waste disposal method that costs the least and gives the most in return. The methods of meeting these criteria will vary from community to community.

● Teacher Notes for "What's Waste?"

This activity gives your students the opportunity to investigate the materials that make up a typical waste stream. They will examine the composition of the waste you provide and consider how alternatives might reduce landfill disposal.

Group Size .... 4–5 students
Time Required .... Getting Ready: 10 minutes
Procedure: 40 minutes
Cleanup: 15 minutes

Safety and Disposal

Emphasize to your students that anything that has been disposed of in a waste container could potentially be dangerous. Therefore, all students must wear goggles and gloves and exercise caution when handling the trash. Classroom trash cans may contain sharp or biologically hazardous items. Therefore, it is recommended that the bags of trash be brought in from elsewhere and that messy, sharp, or otherwise dangerous items be sorted out in advance and, if possible, replaced with safer items from the same category. (Dangerous items could be summarized or saved to teach students about the dangers of picking through trash.) Seal items such as food waste in a plastic bag or container and tell students to leave them in the bag.

No special disposal procedures are required.

Materials

Per group
• bag of trash
• large plastic sheet or trash bag cut open
• 7 small, clear trash bags
• hand-held scale

 A hand-held scale will be easier to use than standard balances, which are likely to be too small. Groups may share scales if necessary.

Per student
• goggles
• pair of rubber or latex gloves

Getting Ready

If you are planning on using classroom waste for this activity, notify your custodian not to empty the cans for a couple of days prior to the activity. This will give students the opportunity to experience a "garbage strike" by going without the convenience of daily trash removal. You or another adult should sort through the trash carefully to ensure that it contains no dangerous items.

If you are bringing waste in from home, distribute an even mix of items among the bags. Also, leave recyclable and compostable items in with the other waste even though you may be in the habit of removing them. This will give students a better representation of the municipal solid waste stream. Set the bags of trash next to the regular trash cans to give the illusion of a "garbage strike." If students ask about the excess trash, you may wish to pretend to know nothing about it until you are ready to begin the activity.

Whatever the source of the waste, make sure that the waste contains materials from as many EPA categories as possible. Also, in order to provide a more true-to-life sample, make sure that no more than 50% of the waste belongs in any one EPA municipal waste category.

Opening Strategy

1. Ask if students notice that the trash can seems to be fuller than usual. Whose responsibility is it to take out the trash at home? How many containers of trash go out to the curb every week? Do they know where it goes? What would we do if no one took the trash away?

2. Introduce students to the concept of the solid waste stream. Explain that waste that must be disposed of is generated daily.

3. Ask students to name the types of things that people throw away and write the list on the board. Once this list is generated, ask students to estimate what percentage of the total trash each category represents. Would their estimates change for trash cans located in different rooms, buildings, homes, businesses, or factories? Would their estimates change at different times of year, such as during the holidays or the leaf-raking season?

4. Distribute the materials to each group and the Student Instructions to each student. Assign items from the list on the board to categories of waste listed in the Classroom Waste Characterization table in the Data Recording section. As a class, discuss the categories and, if possible, come to a consensus as to where

each item fits. Discuss the potential danger of some of the items in the waste stream, review the safety information for the activity, and have the groups begin working.

You may want to bring the groups together after steps 7 and 11 of their instructions to discuss results and conclusions as a class.

Extension

Have students measure the mass or record the number of bags of trash that are thrown out in their households each day. Do the masses or amounts change from day to day? Have them calculate the daily average after keeping track of trash mass for a week or a month. What is the daily average? Have students keep a waste diary in which they record all of the items their family throws away for a week, month, or other time period. Their records should include trash resulting from meal preparation at home.

Cross-Curricular Integration

Art:

• Have students think of creative uses for some of the items in the trash. Provide them with clean "trash" items such as Styrofoam™ cups, plastic cups, coffee cans, and soft-drink cans, as well as pens, crayons, tape, scissors, construction paper, and pipe cleaners. Instruct them to make trash "bugs" using these items.

Mathematics:

• Using the estimated amount of garbage produced per day, calculate how many bags are produced per school year in your classroom and in all of the classrooms combined. Add in estimates for the offices and cafeteria and come up with an estimated annual school-wide total.

Explanation

Before considering different waste-stream management options for a particular community, it is important to determine the nature of the waste stream. For example, waste that consists predominantly of food materials would be a better candidate for composting than waste that contains high amounts of plastic and paper. We affect the waste stream with the products we choose to consume and the decisions we make about disposing of those products. Analyzing the waste stream can help us target items that can be reused or recycled, or that are unnecessary.

● Student Instructions for "What's Waste?"

This activity gives you the opportunity to investigate the materials that make up a typical waste stream. You will examine the composition of classroom refuse (or other waste provided by your teacher) and consider how alternatives might reduce the quantity of waste that must be landfilled.

Safety and Disposal

Anything that has been disposed of in a trash container could potentially be dangerous. Therefore, wear goggles and gloves and exercise caution when handling the trash.

No special disposal procedures are required.

Procedure

1. Measure the mass of the entire bag of trash provided by your teacher. Record the mass in the Data Recording section of this handout. Be sure to include the unit of measure along with the number.

2. Spread the plastic sheet on the floor. Have one member of the group carefully empty the bag of waste onto the sheet, being sure to spread the contents out so the whole group can see all the items. Remember to wear goggles and gloves when you handle the trash.

3. Classify each of the items from the bag according to one of the waste stream categories in the Classroom Waste Characterization table in the Data Recording section. List each item in the appropriate row in the table.

4. Put the waste items from each category into a separate clear trash bag. Measure the mass of each bag of categorized waste and record it in the Classroom Waste Characterization table.

5. Calculate the percentage by mass of each category of waste in your sample and record it in the table.

6. Plot your percentage data in the blank columns (marked "Class") of the EPA Data and Classroom Findings bar graph.

7. Compare the columns for each category of waste. Discuss within your group why the composition of the waste stream you investigated may differ from the EPA's figures. As a group, be prepared to share your results and discussion with the class.

8. Answer questions a–d in the Analysis Questions section of this handout.

9. Suppose that the community producing your waste also has reuse, recycling, and composting programs. Remove all the items covered by these programs from the various bags, return them to the original waste bag, and determine their mass. Record the mass in the Data Recording section.

10. Dispose of all of your bags of waste in a trash can.

11. Calculate the percent of the total mass of trash represented by the materials you removed in step 9. Record the percentage and answer question e in the Analysis Questions section.

Data Recording

_____   Initial total mass of bag

Table: Classroom Waste Characterization

Waste Stream Category	Items in Waste Sample	Mass of Waste	Percent of Total Mass
paper			
plastic			
glass			
metals			
yard wastes			
food wastes			
other			

_____   Mass of items removed for reuse, recycling, or composting

_____   Total percent by mass of all reusable, recyclable, and compostable items removed. Use the following formula:

$$\frac{\textit{Mass of items removed for reuse, recycling, or composting}}{\textit{Initial total mass of bag}}$$

Bar graph: EPA Data* and Classroom Findings

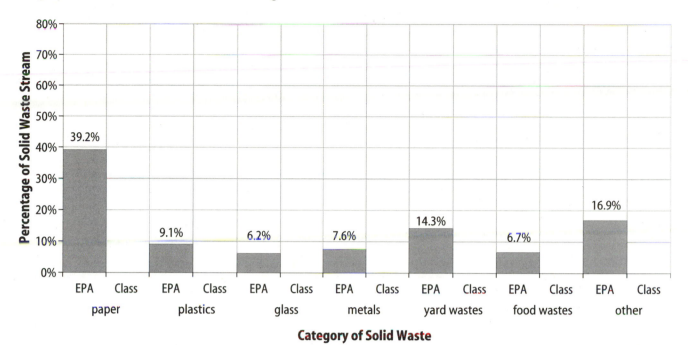

* Data from EPA, 1995

Analysis Questions

a. Could any items have been repaired or reused instead of thrown away?

b. Which items could be recycled?

c. Which items could be composted?

d. Which of the waste stream categories are most likely to contain items that could be removed from the waste stream prior to landfilling?

_____

_____

_____

e. How does the removal of items for reuse, recycling, and composting affect the waste stream?

_____

_____

_____

The Garbage Gazette

October 26   Local Edition   Vol. 1, Issue 2

Garbage Takes a Vacation

Are you packed for vacation? Got your toothbrush? Clothes? 3,100 tons of garbage?

You may not consider 620,000 pounds of solid waste a trip necessity, but the crew of the tugboat *Break of Dawn* wouldn't have left on March 22, 1987, without it. The tug was hired to transport the *Mobro 4000* barge carrying solid waste from New York City to a gas conversion plant in Morehead City, North Carolina. Instead, it spent 55 days traveling around the Atlantic Ocean before returning home, cargo on board.

The history of this infamous cruise begins in the suburbs of New York City. Landfills around the city were being forced to close, and a few suburbs hired a private company to find other places to dispose of their solid waste. This company made arrangements to send the waste to the North Carolina plant.

When the tugboat arrived at the port, the crew was told that necessary paperwork had not been submitted and they would not be allowed to dock. They headed south, going as far as Mexico, Cuba, and Belize without finding anyone who would accept their cargo. During the journey, specialists continually tested the barge to ensure that its contents were not leaking into the ocean.

Finally, the tugboat headed home, where it was not allowed to dock, either. The company who had made the original arrangements was

The *Mobro 4000* garbage barge became a symbol of the American "garbage crisis." (Photo courtesy of John Conover, Jr., P.E.)

bankrupt, and the garbage barge was without a home. Three months later, it still sat, fully loaded, in New York Harbor. Finally, the state of New York intervened and devised a solution. The barge's cargo was incinerated in September, 1987, and the ash was buried in a landfill in Islip, New York.

The garbage barge stole the headlines in 1987 as a symbol of the American "garbage crisis." People believed that the barge was turned away because there was no room in any landfills. However, there was no "crisis." Landfills in other states and countries had room, but since proper procedures were not followed, they could not dispose of the cargo. Had the proper paperwork been submitted originally, the gas conversion plant would have accepted the cargo and the whole incident would have been avoided.

Even though the garbage barge journey was blown out of proportion and misunderstood, the event served a positive purpose. Since 1987, Americans have become more

aware of solid waste issues. Recycling rates have risen, and less garbage per person reaches landfills. Though there is currently no landfill shortage nationwide, the heavily populated East Coast is one area where available landfill space is very limited. Careful management will help avoid a time when barges like the *Mobro 4000* really have nowhere to put the garbage.

Think About It

1. *Where did the barge leave from? Where was it supposed to go? Where did it actually go? Use a map or atlas to determine about how far it was supposed to travel and how far it actually traveled.*

2. *If they can't dispose of their waste nearby, where are other places cities on the East Coast can dispose of it?*

3. *Besides a landfill and recycling, what are some other options for disposing of solid waste?*

The Garbage Gazette

November 7   Local Edition   Vol. 1, Issue 3

Archaeologists Find Garbage Tells the Truth

Is it true that archaeology is rubbish? Golden statues may be good material for an Indiana Jones movie, but archaeologists often gain more information about the lifestyles of ancient people from ordinary items left behind—in other words, ancient garbage. But a group of archaeologists with the University of Arizona's Garbage Project is skipping both the golden statues and the ancient garbage. They're going straight to modern-day trash cans and landfills to learn about the modern American lifestyle.

Over a period of 20 years, these archaeologists have collected, sorted, and identified items contained in 1,766 cubic yards of garbage. Items then are placed into larger categories representing food, drugs, personal and household sanitation products, and materials related to entertainment, education, communications, pets, and yard.

By studying data collected from garbage, archaeologists discover different things, such as how much edible food is thrown away, what percentage of newspapers, cans, bottles, and other items are not recycled; how loyal consumers are to brand-name products; and how much household hazardous waste ends up in landfills.

By interviewing people whose garbage they collect, archaeologists also have discovered that human perceptions about garbage are often

Do you think this pile of garbage can teach us something?

very different from reality. For example, people give inaccurate estimates of how much garbage they produce, how much food they eat, and how much food they waste.

Many Americans think fast-food packaging and disposable diapers comprise a significant percentage of landfill waste; however, together these materials comprise less than 2 percent of the total volume of the average landfill's total solid waste contents. In contrast, paper and debris from construction and demolition account for over half of America's refuse.

Scientists with the Garbage Project believe, "If our garbage, in the eyes of the future, is destined to hold a key to the past, then surely it already holds a key to the present" (Rathje and Murphy 1992, p. 11). What story do you think your garbage would tell about you?

Think About It

1. What do archaeologists study and why?

2. What materials account for over half of America's landfill waste? Does this surprise you? Why?

3. If archaeologists studied the things your family threw away this week, what would they learn about you?

4. What are some ways you can reduce the amount of garbage your family produces?

Lesson 3:
Health Concerns

In this section, students investigate the connection between solid waste disposal and human health. The demonstration "**You *Can* Catch Me, You Dirty Rat**" is a contagion simulation that shows students how easily they can be exposed to pathogens. During the interactive game "**Name That Disease**," students learn what causes diseases, how they are transmitted, and what role solid waste disposal plays in allowing these diseases to spread.

The Garbage Gazette "**A Thousand Years Without a Bath**" presents changing attitudes about hygiene throughout history and epidemics caused by poor hygiene. "**Toxic Waste and You**," the second *Garbage Gazette,* explores the story of Love Canal and the issues involved in health research.

● Teacher Background on Health Concerns

Ancient literature contains references to diseases of all kinds. Some early theories of disease revolved around religion: for example, many people believed disease was a punishment from the gods for some offense. The fact that people who lived in the dirtiest parts of town, which were usually full of garbage and other wastes, were sick more often than those in cleaner areas merely seemed to be consistent with this theory: the poor (who lived in dirtier areas) were not as good or holy as the rich (who lived in cleaner areas). The difference in disease rates among different classes of people is not surprising in the light of modern knowledge of disease. The urban poor lived close together, making it easier to transfer diseases from person to person; had no waste management system to eliminate the insects, rats, and other disease-carrying organisms (called vectors) that lived near garbage piles; and did not have access to clean water. As cities became more crowded, epidemics—events in which diseases affect a large number of individuals within a region at the same time—began affecting rich and poor alike and the theory of disease as divine punishment began to lose credibility.

People then turned to mythology to explain the phenomenon. A popular theory in the Middle Ages was that foul-smelling air or dragons and other serpents caused disease. The serpents that caused disease were supposed to live in pools of stagnant water and have foul breath that caused anyone who inhaled it to be ill. Actually, one element of this theory—stagnant water—is supported by modern scientific research in the case of at least one major disease: malaria, whose name comes from "mal aria," the Italian phrase for "bad air," is transmitted by mosquitoes that breed in stagnant water.

Other theories developed, but it took many years to find the true cause of some diseases. One reason is that many epidemic diseases have similar symptoms, such as fever, rash, nausea, and respiratory problems. Since people didn't know until the 18th century that bacteria, viruses, and other microscopic organisms even existed, they did not know to look for the unique pathogens that caused each disease. Also, in translating ancient texts, scholars used the word "plague" to describe any disease mentioned, regardless of whether the symptoms matched those of the specific disease now known as the plague. Researchers who used these translations as a basis for tracking epidemics through history thus could not draw accurate conclusions about the identities of various diseases.

Understanding Garbage and Our Environment

Another reason for the difficulty in understanding disease was that some epidemics seemed to hit only certain people, most of them children. For a long time, no one understood why this happened. These diseases were called "childhood" diseases, and they struck in 7- to 15-year cycles, affecting mostly children who had not been alive during the last cycle. We now understand that few adults contracted the diseases during subsequent cycles because they had been exposed to the pathogen during an earlier cycle and had built up immunity. A person who has become immune is not affected by the identical pathogen a second time. Some diseases, such as the common cold, may seem to defy this principle, but actually they do not. More than 200 kinds of cold viruses exist, so once infected by one virus, people are immune to just that one; they can still be infected by another.

People gradually build up immunity to the pathogens in their geographic area, so epidemics historically occurred when new pathogens were introduced. At first, this happened by natural causes such as flooding, windstorms, and migration. Then, as irrigation techniques developed, people brought water and the pathogens in it to their farms. When people began traveling to more places, they carried their regional pathogens with them and brought new ones home. Diseases then affected more people, and pandemics became possible. A pandemic is an epidemic that affects more than one continent. Most pandemic diseases were introduced to a continent through a port city, the main point of contact between cultures prior to the 1800s, and spread inland along trade routes.

During the 18th and 19th centuries, disease research advanced significantly, and disease prevention measures were undertaken on a large scale in cities. Waste management became a civic responsibility instead of a private, personal endeavor. Cities hired people to remove the garbage and wastes from the cities and take it to the outskirts of town every day or every few days. Water was treated before it was given to the public to consume. People began taking baths and washing regularly.

The germ theory of disease also began to gain acceptance during this time. Scientists such as Louis Pasteur and Robert Koch advanced the theory that tiny living organisms invisible to the unaided human eye affected human health. The microscope, which had been invented in the late 1500s, began to be widely used as scientists started looking for the tiny "germs" that caused disease.

Once "germs" were found and identified, scientists could start looking for ways to prevent and cure diseases. They could also distinguish for certain between diseases with similar symptoms by looking for specific pathogens. When they determined which vector spread a disease, they could take steps to eliminate the vector and its ability to multiply. Two of the most effective measures were to purify water supplies and remove garbage from the cities. Another major advance in disease prevention was the development of vaccines for certain diseases.

While technology exists to prevent many of the epidemic diseases caused by unhygienic conditions, unfortunately these diseases are still a major problem in many parts of the world. Water purification and waste management systems are nonexistent or inadequate in many countries around the world. Vaccines exist for many of the known major epidemic diseases, but not everyone has access to them. In these situations, lack of money is more of an obstacle to disease prevention than lack of knowledge is.

● Teacher Notes for "You *Can* Catch Me, You Dirty Rat"

This teacher demonstration introduces the concept of vectors in disease transmission and shows students how easy it is to be exposed to a vector.

Group Size .... Demonstration
Time Required .... Getting Ready: 10–15 minutes
Procedure: 15 minutes
Cleanup: 10 minutes

Materials

For Getting Ready
Per demonstration
- 4 rubber bands
- 2 cotton balls
- 2 pencils
- 6 cotton-tipped swabs
- cup large enough to dip "rat feet" into
- liquid detergent that whitens without chlorine bleach, such as Tide®
- meterstick or other long, narrow stick
- tape

For the Procedure
Per class
- black light (UV light)
- extension cord

Safety and Disposal

Only the teacher should operate the UV light. Avoid shining it directly into the students' eyes. Reuse or discard the "rat feet" after making the tracks. Wipe down the "rat tracks" with a wet cloth after the demonstration. Have all students wash their hands thoroughly after the demonstration to remove any residual detergent.

Getting Ready

1. To make a "rat foot," use a rubber band to secure a cotton ball over the eraser end of a pencil. This is the pad of the rat's foot. (See Figure 1a.) Line up three cotton-tipped swabs (the toes of the rat's foot) side by side along the pencil so that all of their tips are even with the cotton-ball-covered end. Secure the swabs in place with another rubber band, arranging the cotton swabs to form three "toes" on one side of the "foot." (See Figure 1b.) Repeat this procedure to make a second "foot" with the other set of materials.

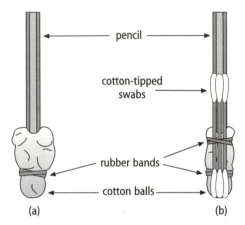

Figure 1: Make a "rat foot."

2. Before the class arrives, make "rat tracks" around the room as follows: Pour a small amount of the liquid detergent into the cup, then dip the rat feet into the detergent. Touch the rat feet alternately along classroom surfaces to create "tracks," redipping the feet in detergent as necessary. Be sure to make tracks across highly used areas such as the pencil sharpener, faucet handles, backs of chairs, desk tops, and other surfaces students are likely to touch several times during the day. To make the demonstration realistic, be sure to run the "rat" back and forth to the classroom trash cans several times. To make prints on the floor, tape each "rat foot" to a meterstick or other long stick so you will not need to stoop or crawl on the floor.

Opening Strategy

Tell students to imagine that a problem with litter and improper waste disposal has developed near the school. What would be some signs that this was occurring? *Odors, flies, wild animals.* What kinds of animals hang around garbage? *Rats, raccoons, dogs.* What would happen if a rat got into the classroom? What kinds of problems could it cause?

Procedure

1. Introduce the concept of a vector: an organism or substance that transports disease-causing bacteria, viruses, and other microorganisms (called pathogens) from one place to another. Examples of vectors include insects, food and water, animals, and people. Explain that many vectors often go unnoticed or unrecognized. The cool, refreshing stream may look like an inviting place for a drink of water, but it could contain microscopic cholera bacteria. Things in the classroom may look normal, but a rat could have found its way in during the night to nibble on lunch leftovers that were not thrown away properly. That rat could carry a hantavirus or fleas that carry *Yersinia pestis,* the pathogen that causes the plague. Not all food, water, insects, animals, and people are vectors, but often you can't tell just by looking if something is a vector or not.

2. Ask students if they see any evidence that a rat has been in the classroom. What kinds of clues are they looking for? Can they find any evidence?

3. Tell the students that a rat walked through some spilled laundry detergent near the school dumpster and that this detergent has a special chemical that makes it glow under black, or UV, light. Explain that you are going to use such a light to see if that same rat has been in the room. Plug the black light into the extension cord. Turn off the classroom lights and turn on the black light, illuminating various parts of the room. Make sure you hold the light near the surfaces where the "rat" walked. Follow all the tracks, then turn the lights back on and the black light off.

4. Ask if students have touched or gone near any of the surfaces the rat walked across. If they have, they may have come in contact with any pathogens carried by the rat or its fleas and ticks. Do students remember all of the surfaces they have touched since entering the classroom? Did they touch something or hand it to another student? Did anyone sneeze or cough? These are all ways that a person can transmit pathogens to other people. Between students going near or touching places where the rat has been and students interacting with each other, you may find that all students in the class could have been exposed to any pathogens the rat carried.

5. Explain to students that there was no rat in the room and that you made the tracks yourself to illustrate how easy it is to come in contact with a possible vector without realizing it. Few rodent-borne pathogens are hardy enough to be transmitted merely by touching a place a rat has walked across. However, hantaviruses are airborne pathogens released in the urine, saliva, and droppings of contaminated rodents. Not all rodents carry hantaviruses, but people can

inhale these pathogens if they are near a place where a carrier has urinated. Also, many parasitic vectors such as fleas and ticks that feed on animals' blood will feed on human blood if the opportunity arises, thus exposing people to any pathogens the parasites may carry. Some human-borne pathogens are airborne, as well, including the pathogens that cause colds and flu. These pathogens are released when people cough or sneeze and can be inhaled by other people. Chicken pox and other rashes can be spread by human contact.

6. Discuss with students how they can minimize contact with animal and insect vectors. If someone has a cold, the flu, or chicken pox, what can they do to help prevent passing the pathogens to others?

Extension

Discuss different types of pathogenic transmission. How would disease transmission be different if microorganisms could survive only in water, required constant direct contact with the host, traveled in a parasite, or were airborne?

Cross-Curricular Integration

Life science:

- Tie this activity into a lesson on hygiene. Put a fluorescent dye on your hands or in a place many students will touch, like a pencil sharpener, before the students arrive and go about your normal routine. (A dye called Glo-Germ® is available from Flinn Scientific, catalog #AP9080, 800/452-1261 or from the Glo-Germ Company, 800/842-6622.) Before the students leave, use a black light to see how far the dye has spread through the classroom merely by touch. Cold and flu viruses can spread in this way. How many other students have dye on their hands? Explain to students that even if we can't see them, we can prevent some pathogens from entering our bodies and making us sick. Some of the easiest ways to keep from getting sick are to keep our hands clean and to properly store, prepare, and dispose of food. Also, avoid touching any unfamiliar animals, especially stray or wild animals, which might bite or carry fleas and ticks.

Explanation

Just as we rely on vehicles to take us places, pathogens rely on carriers called vectors to move them from one location to another. Pathogens have different methods of traveling from vectors to hosts, the organisms that provide the proper environment and nourishment for pathogenic multiplication. Pathogens enter hosts when those hosts interact with vectors, which can be insects, food or water, or a mammal (including human beings). When an insect bites you to drink your blood, it can inject pathogens into you. Some insect vectors, such as fleas and ticks, are themselves commonly carried by animals, especially rodents. When you drink water from a pond or stream without first purifying the water, you also drink any pathogens that are in the water. Some pathogens are airborne and can be inhaled. Other kinds can be spread by direct contact with contaminated surfaces.

Pathogens rely on unhygienic conditions to multiply and spread, and poor waste management promotes these conditions. Improperly discarded food waste attracts rodents and flies and can cause the populations of these disease vectors to increase. Insects, such as fleas and ticks, may be carried by hungry rodents to sources of improperly discarded food waste. *Yersinia pestis,* the pathogen that causes plague, is carried by rodent-borne fleas. Rodents themselves can pick up pathogens on their feet and track the pathogens around, or they can spread the pathogens through feces, urine, or saliva. Hantavirus can be inhaled by people who are near a place where a carrier has urinated. Certain kinds of waste, such as discarded automobile tires, collect water that stagnates and creates an ideal breeding ground for mosquitoes, which can carry malaria. When people throw organic waste into bodies of water, the water may become contaminated with pathogens. Cholera is spread when untreated human sewage is dumped into a drinking water supply.

Humans themselves can be disease vectors. Coughing and sneezing with the mouth and nose uncovered can release pathogens into the air, where they can be inhaled by other people. The common cold and influenza can be spread in this way. When people cough or sneeze directly into their hands and then touch surfaces or other people, they can contaminate those surfaces. Other people can then pick up these pathogens by touching the surfaces and then touching their mouths (as in eating), eyes, or noses. Hand-washing and the use of tissues can prevent the transfer of pathogens through these channels.

Understanding the ways in which different diseases are transmitted is essential to preventing those diseases. Not all diseases are spread in the same way, and not all pathogens can survive under the same conditions. Some pathogens are aerobic, meaning that they require oxygen to survive. Others, such as the *Clostridium* bacterium that causes gangrene, thrive in an oxygen-free (anaerobic) environment. Some pathogens can survive outside a host or vector for long periods of time (as when they are deposited on a surface), while others die within a few minutes. Some pathogens require a specific type or species of vector, while other kinds of pathogens can be spread in more than one way. In some cases, one species of pathogen can cause different variations of a disease depending on how it enters the body. An example is *Yersinia pestis,* which causes three different variations of plague: pneumonic, septicemic, and bubonic. Some types of pathogens that cause different diseases are actually related to each other. Such is the case with bacteria of the genus *Salmonella.* Some species of *Salmonella* bacteria can cause salmonellosis, a kind of food poisoning, while *Salmonella typhi* causes typhoid (also called typhoid fever).

● Teacher Notes for "Name That Disease"

In this activity, your students will play a variation of the "Guess Who?" game to investigate different infectious diseases and characteristics of these diseases. They will examine the role that solid waste management has on the spread and control of certain infectious diseases.

Group Size .... 2 students
Time Required .... Getting Ready: 60–70 minutes
Procedure: 20–30 minutes

Safety and Disposal

No special safety or disposal procedures are required.

Materials

For Getting Ready
Per class
- large piece of paper
- large dark marker
- 3 different colors of cardstock or office paper
- "Name That Disease" Game Card templates (provided)
- Disease ID Card templates (provided)
- Disease Card Key template (provided)
- scissors
- rubber bands

For the Procedure
Per class
- set of Disease ID Cards prepared in Getting Ready

Per group
- 2 sets of Game Cards (prepared in Getting Ready) in 2 different colors
- Disease Card Key

Getting Ready

Create two columns on the large piece of paper. Label one column "disease" and the other "number in class afflicted."

Photocopy the Disease ID Cards onto one color of paper. Cardstock is preferable because it is stiffer, so the cards will not tear as easily and can be used again. Cut the individual cards apart and fold them in half. Stack this set of cards and fasten the stack with a rubber band. Photocopy the Disease Card Keys (one for each group) onto the same color of paper as the Disease ID Cards and cut them apart.

Photocopy the Game Cards onto two different colors of cardstock or office paper. Each student in the class will receive one set of cards, and each partner in a pair should have a different color of cards to make organization easier. Cut the Game Cards apart and rubber-band each set.

Opening Strategy

1. Show the students the columns on the large piece of paper. Ask them to name some diseases or illnesses they have had. Write down the diseases named in the first column. After you have six or seven on the list, ask how many students in the class have also had the disease and record these numbers next to the appropriate diseases in column 2. Ask the students how they were treated for each disease. Did they stay home from school? How did they know what disease they had? Did they go to a doctor, care center, or hospital? Were any tests made to help the doctor determine what disease they had? Were they given medicines or told to eat special foods? Sometimes, the doctor may not know what disease the patient is suffering from or how they contracted the disease. In this case, a specialist may be called in to help determine the disease. Give students the Student Background Information for "Name That Disease" (provided) and have them read about these disease specialists.

2. Divide the class into pairs. Tell them that they will have a chance to be both an epidemiologist and a field technician as they play "Name That Disease."

3. Have each student take a card from the set of Disease ID Cards. Advise them not to tell anyone what disease card they have or let anyone else see their card. Explain what the symbols on the cards mean and give each pair of students a copy of the Disease Card Key.

 If you have fewer than 30 students, make sure that the Disease ID Card for HPS (hantavirus pulmonary syndrome), the mysterious disease featured in the Student

Background, is included in the cards the students pick from so the class can learn the identity of the mysterious disease.

4. Distribute the sets of Game Cards to the students, giving one set of each color to each pair of students. Choose a color of cards. The students who have the sets of this color will be the first students in each pair to ask questions. Distribute the Student Instructions and have the groups play "Name That Disease." When all groups are finished playing, have each student share with the class information about the disease he or she diagnosed.

After the game is finished, have the "epidemiologist" who diagnosed HPS share his or her information first.

Extension

Show electron microscope photographs of viruses and use microscopes to view slides of bacteria and other microorganisms. You may be able to borrow these materials from a high school biology teacher. Explain that these microorganisms are living things and discuss how they are different from and similar to plants, animals, and insects.

Cross-Curricular Integration

Home, safety, and career:

- Invite a real epidemiologist, a hospital infection control nurse, or another health care professional to speak to the class about his or her job.

Life science:

- Explain to students that some diseases can spread exponentially, which means that instead of each person infecting just one other person, each person can infect many people, who can then infect many more people, and so on. If diseases were spread by linear growth—that is, if each person could pass the disease on to only one other person—sweeping plagues would not have killed such large numbers of people throughout history. Because every sick individual can contaminate several others, who in turn can contaminate several more, diseases can spread very quickly through dense human populations. To demonstrate exponential growth, put two or three pieces of duckweed in a container of water in a sunny part of the room. Duckweed is available through Frey Scientific (#F11025 or #F11026), 100 Paragon Parkway, Mansfield, OH 44903; 888/222-1332. It may also be available from a garden center that carries aquatic plants. Have students estimate how many days it will take for the number of duckweed plants in the water to double. How long will it take for the number to triple? Have students count and record the duckweed population daily. Were their predictions accurate? At what point do their results exceed their estimates? How long does it take for the number to increase to the point where counting becomes difficult? (This will depend on the size of the container.) For large containers, students can count the number of plants in a small section, such as a square centimeter, and multiply by the number of square centimeters covered by the mass of plants.

Mathematics:

- Demonstrate the concept of exponential growth mathematically. Draw a large dot at the top of the chalkboard. This represents the first person to have the disease. Person 1 infects two other people. Draw two more dots a few inches apart below the first dot and connect them to the first dot with a line. These two people will each infect two more, who will each infect two more, and so on. (See Figure 1.) Continue the tree diagram until you run out of room. Count the dots in each generation and write the totals next to them. What do students notice about these totals? *The totals are all calculated by multiplying two by itself various numbers of times.* The numbers represent the powers of two, which are calculated using exponents. When you calculate a number with an exponent *n*, you multiply that number by itself *n* times. For example, $2^3 = 2 \times 2 \times 2$, or 8. The 2 is the base, and the 3 is the exponent.

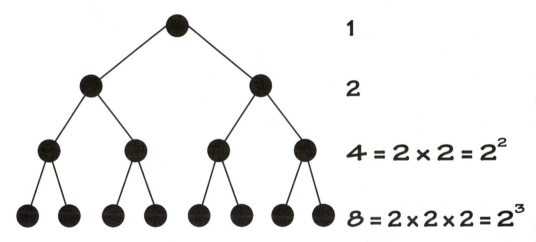

$$1$$
$$2$$
$$4 = 2 \times 2 = 2^2$$
$$8 = 2 \times 2 \times 2 = 2^3$$

Figure1: Use a tree diagram to illustrate exponential growth.

- For more advanced students, do the following demonstration: Assume for the purpose of this demonstration that any person within 1.5 m of a carrier would catch the pathogen and that everyone who carries the pathogen is contagious for 24 hours. (After 24 hours, carriers can no longer transmit the pathogen to others, nor can they be reinfected.) Have every student in the class count how many people he or she would encounter and "infect" over a 24-hour period. Be sure students include their families, other students at school activities and on the bus, their friends, and strangers they may come in contact with in public.

Total these numbers on the board. This number represents the maximum number of people infected after the first 24-hour period. (This number does not take into account the people that come in contact with more than one contagious carrier. If you wish, you can have students devise a way to keep from counting these people more than once in the number used to determine the average infection rate.) Divide this total by the class size to determine the average rate of infection per student in the class. Assume that each of the people infected in the first 24-hour period will then infect the average number of contacts in the next 24-hour period. Multiply the number infected after the first 24-hour period by the average rate of infection to calculate how many new people could be infected at the end of the next 24-hour period. For example, if 250 people were infected in the first 24-hour period and the average rate of infection is 10, then 2,500 people could be infected after the second 24-hour period. Continue to multiply each day's total by the average rate of infection to calculate how many new people could be infected every day for seven days. There are over 260 million people in the United States. At a rate of infection of 10 new people per infected person per day, almost every American citizen could be infected by the eighth day. (See Table 1.) Fortunately, most pathogens are not this infectious, and we do have immune systems, personal hygiene, treatments, quarantines, and vaccinations to help protect the population.

Table 1: Potential Spread of Infection	
After this 24-hour period	This many new people are infected (if each person infects 10 people)
1	25
2	$250 = 25 \times 10^1$
3	$2,500 = 25 \times 10^2$
4	$25,000 = 25 \times 10^3$
5	$250,000 = 25 \times 10^4$
6	$2,500,000 = 25 \times 10^5$
7	$25,000,000 = 25 \times 10^6$
8	$250,000,000 = 25 \times 10^7$

Explanation

Once Louis Pasteur and Robert Koch established the germ theory of disease, effective treatment and prevention of contagious diseases became a reality. Scientists began studying individual pathogens to learn how to keep them out of the body and to destroy them once they entered the body. The "germs" were classified as bacteria, viruses, or other organisms. The majority of contagious diseases are caused by bacteria and viruses, and most epidemiologists concentrate on studying these organisms.

Bacteria

Although they have no nucleus, bacteria have a nuclear region, called a nucleoid, that contains DNA. The average bacterium is about 1 micrometer (1 μm) long and 0.5 μm in diameter (a micrometer is 1×10^{-6} meters, or 0.000001 meters). In comparison, the average human red blood cell is 5–7 μm long, and the average skin cell is 500 μm long. Bacteria come in three basic shapes—coccus, bacillus, and spiral—and some have whip-like appendages called flagella that help them move.

Bacteria are classified as living organisms because they grow and divide. As the one-celled bacterium absorbs nutrients from its environment, it grows larger. When the cell has approximately doubled in size, it divides into two cells, which also grow and eventually divide. The "generation time" of bacteria refers to how long it takes for a single bacterium or bacterial culture (a group of bacteria) to double in number. A bacteria culture containing 100 cells with a generation time of 30 minutes will grow to 200 cells after ½ hour, 400 cells after 1 hour, 800 cells after 1½ hours, and 1,600 cells after 2 hours. Most bacteria have a generation time of 20 minutes to 2 hours.

Viruses

Ranging in size from 10 nanometers to 1 micrometer (a nanometer is 1×10^{-9} meters, or 0.000000001 meters), viruses are even smaller than bacteria, and some viruses, called bacteriophages, infect only bacteria. For example, *Corynebacterium diphtheriae* must be infected with a bacteriophage in order to cause true diphtheria in humans. The bacteriophage contains the genetic codes for the toxins that result in diphtheria. Uninfected *C. diphtheriae* causes a less severe diphtheria-like disease. Unlike bacteria, viruses do not grow and divide. Rather, they replicate only when they are within a living cell. Once a virus gains entrance into a cell, it makes many copies of itself. These copies break out of the host cell, which damages the host cell and may destroy it.

Infection

When a microorganism enters your body, special cells "recognize" that this microorganism is a foreign entity. These cells are able to "remember" which foreign particles have been in your body before. If they do not remember a particle, they begin producing antibodies—special proteins that bind to the foreign organism and prevent it from functioning—until they find an antibody that is effective in stopping the foreign organism. If they do remember the particle, they begin producing the specific antibodies that will stop that organism. In general, cells called B-lymphocytes work against bacteria and T-cells work against viruses, but some overlap does occur.

Because bacteria and viruses begin dividing and replicating once they enter the human body, the B-lymphocytes and T-cells may have thousands of organisms to attack at once. While these cells are searching for the proper antibody, the invading organism continues to multiply. Once the proper antibody is found it may take a few days to destroy all of the invading organisms even though antibodies can be produced quickly—plasma cells, which are specialized B-lymphocytes, can produce 2,000 antibodies per second. Microorganisms that inhibit or destroy B-lymphocytes and T-cells are especially dangerous. For example, the HIV virus, which causes AIDS, destroys T-cells, debilitating the immune system and allowing infections that are normally easily overcome to spread unchecked through the body.

Immunity

Immunity to various pathogens can be attained in different ways. Innate immunity is a genetically determined characteristic. For example, humans are innately immune to canine distemper. They are born immune to it and no matter how many infected dogs they come in contact with, they will not contract the disease. Innate immunity can also be attained through protective physiological features, such as skin. If you are not innately immune to something, you can acquire immunity. Naturally acquired immunity happens after you have been exposed to a pathogen. Your body remembers which antibodies worked against that pathogen, and those same antibodies will destroy the pathogen the next time it is in your body. Artificially acquired immunity occurs when a small amount of the pathogen is injected into your body intentionally. This pathogen is altered just enough so it won't cause the disease but will still cause an immune response. You will still build antibodies, and those antibodies will attack the pathogen the next time it enters your body, but you did not naturally acquire the pathogens the first time. Acquired immunity can also be active or passive. In active immunity, your body generates the antibodies itself, as happens with some vaccines. In passive immunity, ready-made antibodies are injected into the body or are acquired through mother's milk.

Antibiotics and Antivirals

Once scientists began understanding how the immune system worked, they looked for ways they could help the body clear up infections more quickly. The first antibiotics, chemicals that keep bacteria from functioning and eventually kill them, were developed in 1917 after scientists observed that certain bacteria do not grow in the presence of other microorganisms. In 1928, Alexander Fleming noticed that a culture of *Staphylococcus* bacteria would not grow near a colony of *Penicillium* mold. Although this had been noticed by other scientists in the past, Fleming was the first to recognize that this observation could be used to halt infections. Since then, many natural antibiotics have been found and synthetic antibiotics have been developed.

Some antibiotics are bactericidal, which means they actually kill the invading bacteria, and some are bacteriostatic, which means they inhibit the growth of the invading bacteria until you can produce your own antibodies. Regardless of which type of antibiotic a patient is taking, it is important to finish all of the medicine, even if symptoms have subsided. Antibiotics work against the most susceptible bacteria first. When the patient's symptoms go away, the stronger, more resistant bacteria remain, and they are free to multiply in the antibiotic-free environment. The patient's immune system—which can take 1–2 weeks to form a good antibody response—is overwhelmed and cannot destroy all of the bacteria, and the patient has a relapse of symptoms. The original antibiotic will be less effective against the relapse because the bacteria that are multiplying are resistant to it.

Because bacteria are living organisms, they can mutate and change genetically. Later generations of bacteria may not react to antibiotics the same way that earlier generations did, in much the same way that animals and plants have evolved over time. Bacteria can become resistant to certain antibiotics, which means that the antibiotics no longer work against the bacteria. Antibiotics do not create resistant organisms; rather, they create an environment that allows only certain bacteria to survive. These bacteria are ones that have changed enough so that the old antibiotics do not work on them; they no longer have the characteristic that the antibiotic was effective against. In such cases, new tactics are required to treat the disease. For example, scientists have found some bacteriophages that are effective against resistant bacteria. Instead of antibiotics, these special viruses can be used to defeat the bacteria.

Viral diseases are harder to cure than bacterial diseases because antibiotics do not work against viruses. Viral diseases may require the use of antivirals, medicines that inhibit viral replication or prevent viruses from entering a cell but do not destroy the viruses. The antivirals are often extremely toxic to the human body and cause more harm than the pathogen itself. Because they can also mutate with each replication like bacteria, viruses can also become resistant to the antivirals. However, some antivirals have proven effective and relatively safe. Ribavirin, for example, has proven effective against hantaviruses and other viruses in their family.

influenza

Influenzavirus

A Cg
F Fa
L T

The flu is a respiratory disease that rarely causes intestinal problems. One of the largest pandemics occurred during World War I and killed 21 million people. Animals can get the flu, and scientists believe it originated in birds. The virus is passed through droplets ejected while sneezing or coughing and can survive for hours in dried mucus. Because of the virus' rapid mutation rate and many forms, it is almost impossible to create an effective vaccine. Infection with one form does not give immunity from others, so you can get the flu many times.

chicken pox

Varicellovirus

F R

Despite its name, chicken pox is not related to diseases, like smallpox, caused by the family of poxviruses. Rather, it belongs to the family of herpesviruses. Passed through person-to-person contact, chicken pox affects approximately 3.7 million people each year, most of them children. Exposure to chicken pox does not guarantee lifelong immunity, however. Adults over the age of 50 contract the painful skin disease shingles if the virus is reactivated in their bodies. They in turn can give children chicken pox.

cholera

Vibrio cholerae

D V

Cholera is believed to have started in India. Six cholera pandemics occurred between 1817 and 1923. The fourth, from 1863 to 1875 (which struck Europe, the Americas, Africa, China, Japan, and southeast Asia) is geographically one of the largest pandemics in history. Cholera is still a problem in places with poor or no water purification systems. The disease can also be contracted by eating raw or undercooked seafood that has lived in contaminated water. El Tor vibrio, a new strain of cholera that appeared in 1961, is able to travel the oceans in algae.

cold

Rhinovirus

Co Cg
F Fa
T

Over 200 types of the common cold virus exist, which means you can get a cold over 200 times. Pliny the Younger, a Roman who lived in the 1st century AD, prescribed kissing the hairy muzzle of a mouse to cure the cold; if this indeed works, it would be the only known cure. Cold viruses are spread through droplets ejected from a victim while coughing or sneezing and can survive for hours on skin, plastic, wood, Formica®, steel, and most fabrics. Untreated, a cold will run its course in about a week; medications take 7 days to eliminate symptoms.

dengue

Flavivirus

A C
F

Dengue is currently limited to Asia, South and Central Americas, and the Caribbean islands, but anyone visiting these places can contract it. Dengue is rarely fatal and is carried by Aedes mosquitoes, which breed in pools of still, clean water. In 1985, an A. albopictus colony arrived in the U.S. in used, water-filled tires. Although this species carries dengue, the mosquitoes that came to the U.S. luckily were not infected. Dengue has four serotypes. Infection with one does not assure immunity from others, so a person can get dengue four times.

diphtheria

Corynebacterium diphtheriae

Br Cg
S T

Diphtheria, which has existed through most of recorded history, is passed through droplets ejected while coughing or sneezing. The most famous outbreaks occurred in the 1920s in remote Alaskan cities 600 miles away from a source of antitoxin. Blizzards made airplanes and trucks useless, but teams of sled dogs delivered the antitoxin in record time. The annual Iditarod sled race commemorates these heroic journeys. Balto, the lead dog of one of the teams, was honored with a statue in New York City's Central Park, and a movie was made about his journey.

E. coli infections

Your body is used to some strains of *E. coli*, and they have no effect on you when they enter your body. When you ingest a new strain through undercooked meat or contaminated water, the bacteria reproduce at a fast rate—under ideal conditions, one cell could multiply into a mass greater than the Earth in 3 days—and you experience intestinal problems. Flies can spread *E. coli* by walking on contaminated surfaces and then landing on food or water. Because *E. coli* strains vary by region, they are the main cause of intestinal problems in tourists.

Escherichia coli

D F
N V

hemorrhagic fever (Ebola)

Despite medical advances, the Ebola virus is still a mystery. Scientists know it causes a fatal hemorrhagic fever and rolls into a question-mark-like shape for protection; they don't know where it's from and aren't sure of its vector, except that outbreaks seem to follow exposure to a monkey carrying the virus. In 1967, a mild hemorrhagic fever (25% fatal) emerged in Marburg, Germany. Ebola (90% fatal) appeared in Africa 9 years later. Scientists placed these related viruses into a new virus family, of which they are the only known members.

Filovirus

B F
R

hepatitis

Hepatitis is an inflammation of the liver that can be caused by many factors, the most common of which are viruses. A patient is diagnosed with viral hepatitis based on which of five unrelated viruses is infecting the liver. Hepatitis A is spread through contaminated water or undercooked fish from contaminated water. Hepatitis B (pictured) is transmitted through bodily fluids and can survive on needles and surgical tools. Hepatitis C, D, and E are not as common but are also transmitted through bodily fluids and contaminated water.

Hepadnavirus

A F
Fa J

HFRS

Hemorrhagic fever with renal syndrome (HFRS) has been present in Asia for centuries. The disease, caused by a hantavirus, was not widely known in the rest of the world until the 1950s, when 3,500 European and American soldiers were infected during the Korean War. At first, scientists believed the viruses were passed by mites that fed on mice. Later tests showed the disease was passed directly by rodents. Humans contract the disease after inhaling airborne viruses released in rodent urine and saliva.

Hantavirus

A B
C F
N V

HPS

In 1993, a mysterious respiratory disease killed many people in the southwestern United States. None of the pathogens of the usual respiratory diseases were present. However, hantaviruses were present, although the patients showed no symptoms of HFRS, the only known hantaviral disease at this time. Scientists realized they had discovered a new hantaviral disease: hantavirus pulmonary syndrome (HPS). As with HFRS, humans contract HPS after inhaling airborne hantaviruses released in rodent urine and saliva.

Hantavirus

B Br
Cg L

Legionnaires' disease

At a 1976 American Legion convention, 221 people, 34 of whom died, contracted a mysterious pneumonia. The cause: a new bacteria that had grown in air-conditioning cooling towers at the convention site. *Legionella* bacteria (named, like the disease, after the first victims) thrive in wet, dark places and can survive in room-temperature water for a year. Humans contract the disease after inhaling drops of contaminated water dispersed from the source, such as through air-conditioning systems. It is not spread by coughing or sneezing.

Legionella pneumophila

A C
Cg F
Le

Lyme disease

Borelia burgdorferi

A C F Fa R

The first recognized outbreak of Lyme disease was in Old Lyme, Connecticut, in 1975, but new studies show cases occurred in Wisconsin and Massachusetts in the 1960s and in Europe in the 1920s. The bacteria reside harmlessly in white-footed mice. Tick larvae feed on these mice and pick up the bacteria. While the adult ticks prefer deer blood, they will feed on humans, passing on the bacteria to their new hosts. The tick must be attached to a human host for 36–48 hours to infect the host with Lyme disease.

malaria

Plasmodium

C F

Malaria currently affects 400 million people every year, killing 1–3 million people. Egyptian, Biblical, and early Greek writings mention the periodic fevers of malaria. Thought to have started in tropical Africa, malaria now affects most tropical countries. Biologists believe that African people developed sickle-shaped blood cells as protection against malaria. The pathogen reproduces in the body of female Anopheles mosquitoes, who then bite and infect humans (male Anopheles are vegetarians).

measles

Morbillivirus

Co F L R

The most infectious common disease is the measles. American children are vaccinated soon after they are born, but not all places in the world have access to the vaccine. The mucous lining of the respiratory tract provides an ideal breeding ground for the measles virus, which is passed through sneezing or coughing. The earliest record of measles was by the Persian physician al-Rhazes in the 1st century AD. He believed that measles was a natural childhood episode, like losing baby teeth.

meningococcus meningitis

Neisseria meningitidis

A F R V

Inflammation of the meninges (the membranes covering the brain and spinal cord) is caused by many factors, including *Neisseria meningitidis*. Human beings are the only natural hosts in which the bacteria, which are transmitted through nasal droplets, cause disease. The bacteria enter the body through the nasal passages at the back of the throat but do not cause meningitis unless they enter the bloodstream. About 30% of people during an epidemic may show no symptoms but are still able to carry the pathogen and pass it on.

mumps

Paramyxovirus

F G

The mumps is a less dangerous and contagious cousin of the measles. Like the measles, it is spread person to person via droplets ejected while coughing or sneezing. It also lives and grows in the respiratory tract, where it causes the salivary glands to swell. Hippocrates first mentioned the mumps in Greece in about 400 B.C. Approximately 30% of all people who have the mumps virus show no symptoms. Children in the U.S. are given MMR inoculations soon after they are born to protect them against measles, mumps, and rubella (German measles).

plague

Yersinia pestis

B Br Cg F G L S

The largest plague pandemic was the "Black Death" of the 1300s, when a third of the European population was killed and raiders catapulted corpses of plague victims into cities. There are 3 types of plague: bubonic (swelling and bleeding), pneumonic (pneumonia-like symptoms), and septicemic (fatal blood poisoning). Bubonic and septicemic plagues are spread by rat fleas, and pneumonic plague is spread through sneezing or coughing by victims. The type of plague a person gets depends upon where in the body the bacteria multiply.

polio

Enterovirus

Poliomyelitis, or polio, has probably been around since ancient times. U.S. president Franklin Delano Roosevelt had polio, and his experiences helped improve the way handicapped people were viewed and treated. Jonas Salk and Albert Sabin created polio vaccines in the 1900s. Children born in the U.S. and most of the world are vaccinated against polio at birth. Although polio is close to becoming eradicated, it still afflicts people without access to the vaccine and with inadequate water treatment facilities.

rubella

Rubivirus

The virus that causes rubella, or the "German measles," is not related to the measles virus, although both viruses cause similar symptoms and are transmitted through coughing or sneezing. Although rubella is not harmful to most people, pregnant women who are exposed to the virus during the first trimester of pregnancy have a higher risk of bearing children with heart problems, deafness, and language disorders. A vaccine was developed in 1969 and is given to U.S. children soon after they are born as part of the MMR vaccine.

salmonellosis

Salmonella

Salmonellosis is caused when a person eats undercooked eggs or meat infected with Salmonella. The disease can also be spread through the handling of green turtles, which is why these once-popular pets are no longer sold in pet stores. The 2,200 kinds of Salmonella are resistant to many antibiotics and can transfer their resistant characteristics to E. coli and other bacteria. Over 2 million Americans get salmonellosis each year, but symptoms are often mild and some never realize they had anything other than an upset stomach.

sleeping sickness

Trypanosoma gambiense

The two forms of sleeping sickness are limited to separate regions in Africa, the natural home of the vector. Natives of and visitors to these regions can be infected. Tsetse flies bite animals (who are symptom-free reservoirs) or infected people and pick up the pathogens. When these flies eat again, they pass the protozoans to their next host. The Rhodesian form is marked by severe blood poisoning and rapid death. The Gambian form, in which the protozoans invade the brain and spinal cord, usually kills the victim after a few years.

smallpox

Orthopoxvirus

Smallpox was highly contagious: one exhaled droplet from a carrier could contain 1,000 more viruses than were needed to infect another. The earliest record dates from 1122 B.C. in China. The Spanish brought smallpox to the Americas, and the disease killed entire native tribes. Pocahontas died of smallpox in 1617. In the 1700s, Edward Jenner discovered that people who had cowpox, a non-fatal rash, didn't contract smallpox. This knowledge was used to develop vaccines, and smallpox has since been completely eradicated.

strep throat

Streptococcus pyogenes

A red, scratchy, sore throat can have many causes; a Streptococcus bacterium infection is just one of them. "Strep throat" is passed through coughing or sneezing or by drinking from the same glass or using the same utensils as an infected person. Prior to pasteurization, it could be passed through a milk supply from a dairy farmer with the infection. Scarlet fever is a more severe case of strep throat accompanied by a rash and vomiting. Rheumatic fever is often a consequence of the infection. A harmless strain of Streptococcus bacteria gives yogurt its creamy taste.

trichinosis

Trichinosis is caused by parasitic worms that enter the body when a person eats undercooked meat. Since many parasites are needed to cause symptoms, only a small percentage of people are infected enough to have symptoms. A few weeks after a person eats contaminated meat, the worm larvae may move to the tongue, eyes, or muscles of the host. Pork is the most frequent source, but bear or walrus meat can also contain the worms. Animals ingest the worms when they eat contaminated garbage.

Trichinella spiralis

A F
N

tuberculosis

Tuberculosis, or "TB," is an ancient disease that dates back to approximately 4500 B.C. It was mentioned in the "Code of Hammurabi," one of the earliest sets of laws, in 1700 B.C. Infection requires months of close contact, and droplets ejected while coughing, sneezing, talking, and singing can expose others to the disease. Simply letting fresh air blow through an infected person's room will remove about two-thirds of the infectious materials. Cows or other animals can get TB and transmit it to humans through unpasteurized milk. Malnutrition increases the chance of getting TB.

Mycobacterium tuberculosis

Br Cg
Co L

typhoid

Related to salmonella poisoning, typhoid is an intestinal disease contracted by eating contaminated food. During the Boer War, 13,000 British soldiers died of typhoid, while only 8,000 were killed in combat. Human carriers pass along the disease while preparing food or beverages. Fly carriers contaminate food when they land on it. There is no evidence of direct human-to-human or fly-to-human transfer. Even if people don't show symptoms, they can still infect others. "Typhoid Mary" Mallon, a cook, was one such transient carrier, infecting 54 others between 1904 and 1914.

Salmonella typhi

A B
F R

typhus

The first reliable description of typhus was a record from a Salerno monastery in 1083. In the 1400s, an undoubted epidemic occurred when Spanish soldiers brought it back from their campaigns in Cyprus. Many Native American tribes were killed when European explorers arrived, bringing the disease with them. Typhus, also called jail fever, is spread by lice. Epidemics occur during war, famine, and any time people are suddenly crowded together and hygiene is neglected. During many early wars, more soldiers were killed by typhus than by the enemy.

Rickettsia prowazekii

A C
F P
R

whooping cough

About 100 years ago, whooping cough, also known as pertussis or "chin cough," was the major cause of infant deaths. Today, children in America and other countries are immunized against the disease soon after they are born. However, about 40 million cases and 400,000 deaths still occur every year. Most people recover from the disease—characterized by a spastic cough that happens 5–15 times in a row followed by a deep, crowing, whoop-like breath that gives the disease its name—in about 7 weeks. The disease is spread through droplets ejected while coughing.

Bordetella pertussis

Br Cg
V

yellow fever

Yellow fever is believed to have originated in tropical Africa and is common in the tropical Americas. Aedes mosquitoes, the vector of the disease, are not native to America; it is believed they arrived via trading ships. Yellow fever halted the construction of the Panama canal and changed the course of rebellions in the Caribbean islands. Napoleon lost 22,000 of his 25,000 soldiers to yellow fever in Haiti, and this helped convince him to sell the Louisiana Territory to the U.S. Epidemics in the U.S. moved up the Mississippi River or hit eastern ports.

Arbovirus

A F
J N
V

influenza

Influenzavirus

A Cg
F Fa
L T

cold

Rhinovirus

Cg Co
F Fa
T

E. coli infections

Escherichia coli

D F
N V

HFRS

Hantavirus

A B
C F
N V

Lyme disease

Borelia burgdorferi

A C
F Fa
R

chicken pox

Varicellovirus

F R

dengue

Flavivirus

A C
F

hemorrhagic fever (Ebola)

Filovirus

B F
R

HPS

Hantavirus

B Br
Cg L

malaria

Plasmodium

C F

cholera

Vibrio cholerae

D V

diphtheria

Corynebacterium diphtheriae

Br Cg
S T

hepatitis

Hepadnavirus

A F
Fa J

Legionnaires' disease

Legionella pneumophila

A C
Cg F
Le

measles

Morbillivirus

Co F
L R

meningococcus meningitis
Neisseria meningitidis
A F R V

mumps
Paramyxovirus
F G

plague
Yersinia pestis
B Br Cg F G L S

polio
Enterovirus
A F P

rubella
Rubivirus
F G R T

salmonellosis
Salmonella
D F Fa N V

sleeping sickness
Trypanosoma gambiense
F Fa G Le P S

smallpox
Orthopoxvirus
B F R

strep throat
Streptococcus pyogenes
A F N T

trichinosis
Trichinella spiralis
A F N

tuberculosis
Mycobacterium tuberculosis
Br Cg Co L

typhoid
Salmonella typhi
A B F R

typhus
Rickettsia prowazekii
A C F P R

whooping cough
Bordetella pertussis
Br Cg V

yellow fever
Arbovirus
A F J N V

Pathogen

Virus

Bacteria

Other

Vector

Insects

Mammals (including humans)

Food and Water

Waste Mgt. Link

Hygiene

Dumps

Contamination

Symptoms

A aches
B bleeding
Br breathing difficulties
C chills
Cg coughing
Co congestion
D diarrhea

F fever
Fa fatigue
G swollen glands
J jaundice
L lungs fill with fluid
Le lethargy

N nausea
P paralysis or numbness
R rash
S septicemia (blood poisoning)
T sore throat
V vomiting

Pathogen

Virus

Bacteria

Other

Vector

Insects

Mammals (including humans)

Food and Water

Waste Mgt. Link

Hygiene

Dumps

Contamination

Symptoms

A aches
B bleeding
Br breathing difficulties
C chills
Cg coughing
Co congestion
D diarrhea

F fever
Fa fatigue
G swollen glands
J jaundice
L lungs fill with fluid
Le lethargy

N nausea
P paralysis or numbness
R rash
S septicemia (blood poisoning)
T sore throat
V vomiting

Pathogen

Virus

Bacteria

Other

Vector

Insects

Mammals (including humans)

Food and Water

Waste Mgt. Link

Hygiene

Dumps

Contamination

Symptoms

A aches
B bleeding
Br breathing difficulties
C chills
Cg coughing
Co congestion
D diarrhea

F fever
Fa fatigue
G swollen glands
J jaundice
L lungs fill with fluid
Le lethargy

N nausea
P paralysis or numbness
R rash
S septicemia (blood poisoning)
T sore throat
V vomiting

Pathogen

Virus

Bacteria

Other

Vector

Insects

Mammals (including humans)

Food and Water

Waste Mgt. Link

Hygiene

Dumps

Contamination

Symptoms

A aches
B bleeding
Br breathing difficulties
C chills
Cg coughing
Co congestion
D diarrhea

F fever
Fa fatigue
G swollen glands
J jaundice
L lungs fill with fluid
Le lethargy

N nausea
P paralysis or numbness
R rash
S septicemia (blood poisoning)
T sore throat
V vomiting

● Student Background for "Name That Disease"

In May 1993, a young man was traveling with his family to his fiancee's funeral when he suddenly began struggling for breath. He was rushed to the Indian Medical Center in Gallup, New Mexico, but the doctors were unable to save the man because his lungs had filled with fluid. The young man had been experiencing flu-like symptoms, but the doctors could not determine what had caused his lungs to fill with fluid. Unable to adequately explain the cause of death, the doctors called the Office of the Medical Investigator (OMI) to report the death, as was required by state law.

Richard Malone, the OMI deputy medical investigator in Gallup, took the call. As he listened to the case, he was struck by the similarities between this case and the death of a woman only a month before. Malone talked to the young man's family to learn more and discovered that the young man's fiancee had also died under similar circumstances. Malone reported these three cases to New Mexico's state health department as a possible outbreak of an unknown, fatal respiratory illness.

Bruce Tempest, the chief of medicine at the Indian Medical Center, was intrigued by the similarities between these strange deaths and a case an Arizona colleague had told him about. He called this colleague again and learned that another person in Arizona had died under the same mysterious circumstances. Tempest reported what he knew to New Mexico's state health department. As word of the mysterious disease spread, more reports of similar cases poured in from all over the Southwest, causing health officials to call in the experts at tracking down mysterious diseases: epidemiologists.

Epidemiologists are chemists, biologists, pathologists, mathematicians, doctors, coroners, and other people who work to identify and control epidemics—outbreaks of disease that affect many people in a certain region at the same time—and other health emergencies. The Centers for Disease Control and Prevention (CDC), headquartered in Atlanta, Georgia, specializes in epidemiology. CDC's efforts led to the eradication of smallpox worldwide and polio in the United States. It also identified the pathogens responsible for Legionnaire's Disease, its milder cousin Pontiac Fever, and the mysterious deaths in the Southwest.

When epidemiologists learn of an epidemic, they first try to establish how it started. They interview victims and their families to find possible patterns. They look for cases outside of the affected region to see if a victim brought the disease from somewhere else. Tests are done on blood samples to look for pathogens. Many times, epidemiologists learn just as much from what pathogens are not present as they do from the pathogens they find. For example, in the case described above, since epidemiologists found no traces of the usual pathogens that cause lungs to fill with fluid, they concluded that they might be dealing with a previously unknown pathogen or a known pathogen that was suddenly causing new symptoms.

Looking for pathogens is not as easy as you might think. Most of them are so tiny that they cannot be seen with the naked eye. Epidemiologists must use a microscope to search for these tiny organisms. Viruses, which are among the smallest known organisms, cannot be seen with a regular microscope; they can be seen only with an electron microscope. To help you visualize how small pathogens can be, the following figure describes the size of a few of these organisms in terms of the size of a nickel. The measurements are given in metric units, to make comparison easier. Remember that 1 centimeter (cm) = 10 millimeters (mm) =10,000 micrometers (μm) = 10,000,000 millimicrons (mμ).

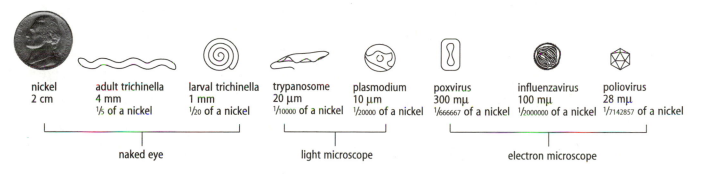

| nickel 2 cm | adult trichinella 4 mm 1/5 of a nickel | larval trichinella 1 mm 1/20 of a nickel | trypanosome 20 μm 1/10000 of a nickel | plasmodium 10 μm 1/20000 of a nickel | poxvirus 300 mμ 1/666667 of a nickel | influenzavirus 100 mμ 1/2000000 of a nickel | poliovirus 28 mμ 1/7142857 of a nickel |

naked eye   light microscope   electron microscope

Once epidemiologists identify the pathogen and its vector (the organism or object that spreads the pathogen), they can better control the disease and help prevent it from recurring in the future. Medications like treatments and vaccines can be developed by studying the pathogen and its life cycle. Epidemiologists also teach people in risk areas to avoid vectors by properly disposing of garbage, practicing proper hygiene, drinking only treated water, and properly preparing food before consumption.

As for the identity of the mysterious pathogen that caused the deaths in the Southwest, you'll find out after you play the game "Name That Disease."

● Student Instructions for "Name That Disease"

A mysterious epidemic has erupted in your town and you have been asked to help determine the identity of the disease. Can you use the critical information your field technician gives you to identify the disease and stop it before it spreads further?

Safety and Disposal

No special safety or disposal procedures are required.

Procedure

This game is based on the "Guess Who?®" Game, with which you may be familiar. A contagious disease has broken out in an imaginary town (you can pick the name of the town). You are an epidemiologist (a person who studies infectious diseases), and you have been asked to identify the disease. You must do this as quickly as possible so you can prevent the disease from spreading outside the town. To make your diagnosis, you must ask "yes or no" questions of your field technician, played by your partner. Your partner is also an epidemiologist asked to diagnose a disease in another imaginary town, and you are his or her field technician. Play the game as follows:

1. Lay your set of Game Cards face up on the table or floor in front of you in a grid pattern. These cards will help you as you try to diagnose which disease is infecting your town. Set your own Disease ID Card (given to you by your teacher) face down nearby so you can refer to it when your partner asks you questions. This card describes the disease your partner is trying to diagnose. Answer your partner's questions based on the information on this card, but do not let him or her see the card. (Your partner has his or her own secret Disease ID Card, which will describe the disease you are trying to diagnose.)

2. Take turns with your partner asking each other "yes or no" questions that will help you determine the following characteristics of the disease:
   • the type of pathogen that causes the disease (virus, bacteria, or other)
   • the vector of the disease (insect, mammal, or food and water)
   • the symptoms of the disease (such as fever, nausea, or chills)
   • the link to solid waste (dumps, hygiene, or contamination)

   Before you ask each question, review the information on the Game Cards in front of you that are still face up. Ask questions that will eliminate the most possibilities.

3. After you have received your field technician's answer, flip over the cards of any diseases that have been eliminated, using the Disease Card Key as a guide to the symbols. For example, if you asked your field technician if the disease was caused by a virus and the answer was "No," you would flip over the cards of all diseases that are caused by viruses, leaving all cards of diseases that are caused by something other than a virus face up. If you asked if a sore throat was a symptom and the answer was "Yes," you would flip over the cards of all diseases that do not have sore throat as a symptom, leaving all cards of diseases that have sore throat as a symptom face up.

4. Continue playing until either you or your partner is ready to guess the name of the disease infecting the town. During your turn, you would guess the name of the disease instead of asking a question. If your guess is correct, your partner may continue to ask questions until he or she also guesses correctly. If the guess is incorrect, the guesser skips a turn and the other person is able to ask two questions in a row. Play continues until you and your partner have both guessed correctly.

5. If you or your partner has one card left but this card does not match the Disease ID Card the other person has, either the field technician gave misinformation or the epidemiologist turned over the wrong cards. If this happens, have the epidemiologist flip over all of his or her cards and begin the game again, paying close attention to the answers given and the cards that are supposed to be eliminated.

6. When the game is over, exchange Disease ID Cards with your partner. Use this card to fill in the information requested in the Data Recording section. Look over this information and answer questions a and b in the Analysis Questions section. You will then use what you have written to tell your class about the disease you diagnosed and how you would prevent it from spreading.

Data Recording

Disease Name:

_____

Name and type of pathogen:

_____

Vector(s):

_____

Symptoms of the disease:

_____

_____

_____

Link(s) to solid waste management:

_____

Interesting facts about the disease:

_____

_____

_____

_____

_____

_____

Analysis Questions

a. What should the residents of your town do to keep this disease from spreading further?

_____

_____

_____

_____

_____

_____

b. What can the residents of the town do to help prevent this or similar contagious diseases from infecting them?

_____

_____

_____

_____

_____

_____

_____

The Garbage Gazette

November 28   Local Edition   Vol. 1, Issue 4

A Thousand Years Without a Bath

A thousand years without a bath? This may seem strange to us now, when showering about seven times a week is average in America, but before people knew that cleanliness could prevent disease, hygiene was treated more as a fashion than a necessity.

Prior to the 4th century AD, most cultures encouraged the daily cleaning of the face and hands, with full-body baths about once a week. The Greeks bathed frequently and would often soak in a tub for brief periods of time to relax. They prized cleanliness but did not let it turn into indulgence. Perhaps the most famous Greek bather was Archimedes, who discovered the principle of water displacement. Stumped by the problem of determining whether a wreath was made of pure gold or of gold and silver, he decided to clear his mind by taking a bath. As he sat down in the water, the water level rose a certain amount. The solution to his problem sprang into his mind and he allegedly sprang from the bath and ran naked through the streets shouting "Eureka!" (Greek for "I have found it!"), although this part of the story is thought to be a myth.

The Romans used their public baths as an arena for discussion and relaxation, but they often went to extremes, spending hours debating issues or discussing news. Due to the popularity of baths, the average daily water consumption was 300 gallons per person, about four times what the average American uses today.

Could you go 1 year without a bath?

A movement that started in Europe in 300 AD succeeded in banning bathing because people believed that it encouraged sinful behavior. This ban continued through the Middle Ages (500–1500 AD), and it's said that Europe didn't take a bath for those 1,000 years. Queen Isabella of Portugal, the sponsor of Columbus's voyage to the Americas, boasted that she had taken only two baths in her life, once at birth and once before her wedding. Some Europeans bathed more frequently: monks, who had access to Greek and Roman writings about the therapeutic value of baths, took 3–4 baths a year.

People at this time also believed that all creatures had a right to life. Veterinary hospitals cared for sick rats and hired people to be hosts for fleas, ticks, and lice! At the other end of the spectrum, people who wouldn't tolerate insects picked them off their bodies and crushed them between their fingers or teeth, which spread insect-borne diseases like typhus and the plague.

The Middle Ages was a time of epidemics such as leprosy, bubonic plague, smallpox, and tuberculosis that could have all been prevented by observing simple hygiene practices like bathing and avoiding rodents and insects. The hygienic practice of quarantining, or isolating the sick, did originate in the Middle Ages, however, in response to contagious epidemics. Separating the sick from the healthy was the only weapon people knew of at the time for combating these diseases. Hospitals created isolation wards, and ships were quarantined at the docks for a period of time before sailors were allowed to disembark. Any sailor showing signs of illness during the quarantine was not allowed to leave the ship. This prevented people from spreading illness, but rodent stowaways that also carried diseases were free to come and go from the ships.

Not all cultures scorned hygiene. Egyptian priests shaved their bodies to keep themselves free of fleas and lice, and cats were revered because they killed rodents. Also in Egypt, fragrant oils and ointments were applied every day, and archaeologists believe these worked like the deodorants and perfumes used today. Manchurian nomads avoided contact with sick marmots (a type of burrowing rodent). Mongol invaders ignored this practice, however, and they contracted plague from fleas that infected marmots. When the invaders moved on to Europe in the 1300s, they carried the disease with them.

The Middle Ages were followed by the Renaissance period in Europe, which is usually considered an age of enlightenment. Advances were made in science and medicine, and scientists began using formal classification systems, making the recognition of diseases more precise. Educated leaders recognized that political and economic strength required a healthy population, and that meant a clean population. Sanitation and hygiene, however, were still in the private realm; the governments of the day lacked the resources to create and maintain such policies. Any hygienic or sanitary practices were implemented and encouraged by private hospitals, charities, and churches but were not always followed by the general public.

Religion continued to be the main influence on hygiene in the 1700s. Washing was permitted, but frequent bathing was forbidden. For example, in colonial Philadelphia, it was against the law to bathe more than once a month. Anyone caught breaking this law faced fines and jail time. The Philadelphians eventually learned their lesson, though. During the flu epidemic of the early 1900s, it was against the law to sneeze or cough in public without a handkerchief.

In the 1800s, hygiene became a civic responsibility as public health engineering—in which garbage collection and sanitation are managed by government—developed and was advanced by many groups, including the ultra-sanitarians. This group, led by Edwin Chadwick, believed that disease was caused by dirt and decomposing matter since diseases usually concentrated in the dirtiest parts of town. Their solution was to clean and ventilate these areas regularly. Private fever hospitals were created for sick people who could not afford to clean and ventilate their own homes. Although only a small group believed Chadwick's theory, these people were very influential and helped establish sanitation as a civic responsibility in Britain.

When Louis Pasteur proved that microorganisms caused diseases, some people began to take hygiene more seriously, but it took years for

Soap helps to protect us from disease.

hygienic practices to catch on. Doctors who didn't believe Pasteur went from doing autopsies to delivering babies or tending to patients without washing their hands. Needless to say, many people died from infections.

Ignaz Philipp Semmelweis, an Austrian doctor, first recognized the connection between childbirth fever and autopsies in 1844 and insisted that doctors wash their hands before seeing each patient. Incidences of childbirth fever dramatically declined in clinics run by Semmelweis. Younger doctors were quick to follow his example, but Semmelweis's peers and superiors—who believed diseases were unpreventable—laughed at him and shunned him, keeping him from getting jobs. He had a nervous breakdown and died in a mental hospital—ironically of an infection by the same microorganism that caused childbirth fever.

After reading Semmelweis's works, English doctor Joseph Lister continued promoting handwashing and started the practice of soaking instruments and bandages in carbolic acid to reduce infection. Despite ridicule, Lister—after whom Listerine® antiseptic mouthwash was named—continued to praise the late Semmelweis's work and credited him as the inspiration for his ideas. Lister's and Semmelweis's ideas gained widespread acceptance in the scientific and medical communities as the germ theory also grew in popularity. By the turn of the century, hygiene was back in style—permanently.

Think About It

1. What is hygiene? What are some examples of hygienic practices?

2. Why is hygiene important?

3. Why do you think early religious and scientific beliefs were more important to people in the past than hygiene?

The Garbage Gazette

December 8   Local Edition   Vol. 1, Issue 5

Toxic Waste and You

Every day, you are exposed to high doses of a potentially lethal compound known as dihydrogen monoxide. This compound can be fatal in small doses and it stays in your body for years. It's present in the foods and beverages you eat and drink, and scientists have yet to find a way to keep this compound out of your body.

How do you feel right now? Confused? Angry? Afraid? What questions do you have? What is dihydrogen monoxide, and why can't scientists keep it away from you?

Because they don't want to. "Dihydrogen monoxide," which consists of two hydrogen atoms and one oxygen atom, is H_2O, more commonly known as water. While it can be fatal if it gets into your lungs, water is mostly harmless. How do you feel now?

When you didn't know that "dihydrogen monoxide" was really just water, you probably felt the same way residents of the Love Canal area in Niagara Falls, New York, felt when chemicals from the dump their neighborhood was built on started seeping into their lives. What were these chemicals? Were they dangerous? Could they be causing unexplained health problems some residents had?

Because of the answers to that last question, Love Canal became a symbol of the American toxic waste issue. But what does it symbolize — the danger of toxic waste or the danger of rash decision-making?

Hazardous waste is an issue many people feel strongly about.

In 1942, the Hooker Chemical Company obtained permission to use an abandoned, clay-lined canal, originally built by William Love in the early 1900s, as a chemical waste dump. Hooker placed drums containing 21,800 tons of chemicals into the canal until they closed it in 1953.

Later that year, Hooker was pressured into selling the site for $1 to the Niagara Falls Board of Education, who knew what the site had been used for. Hooker warned the board not to dig near the canal, as this would disturb the clay cap covering the chemicals. Despite warnings, the board built a school on the canal and the city developed neighborhoods in the area.

Abnormally heavy rains caused the canal to overflow in 1976. Reports of nauseous odors, black sludge, and children being burnt by chemicals near the canal spread through the city. A series of articles in a local paper brought the history of the chemical waste dump to the attention of residents, who understandably started to worry.

The homes near the canal were tested to find out what was seeping from the canal and how far it was going. Low levels of 80 chemicals—including one known human carcinogen (a cancer-causing agent) and 11 known or suspected animal carcinogens—were detected.

When residents discovered that chemicals had been invading their homes, they began to wonder if the chemicals could be responsible for unexplained health problems some of them had. If the doctors couldn't explain why a child had epilepsy, why a man had cancer, why a woman miscarried, or why a baby was born with defects, then the chemicals had to be to blame. What residents didn't realize is that doctors often have no explanations for these problems, whether the patients live near toxic waste dumps or not.

Having no other explanation, some residents blamed the chemicals for health problems and any other "inexplicable" problems like sick animals, dying plants, and sinking or broken sidewalks and porches. They demanded that the school be closed immediately and that the government pay for their relocation to a safer neighborhood. However, before the government could justify any decisions involving Love Canal, they first needed to find out whether the chemicals were in fact responsible for the residents' health problems. Officials arranged for scientific testing to be done.

The Garbage Gazette

December 8   Local Edition   Vol. 1, Issue 5

When scientists try to establish cause-and-effect relationships, they use the mathematical field of statistics, which provides tools for proving hypotheses. In order for statistical data to be meaningful, research must meet certain conditions. In the early tests on Love Canal residents, many of these conditions were not met, which cast serious doubt on the conclusions researchers made.

Every good scientific experiment must have a "control"—something to compare results with. In the case of research on health problems, the control group is a group of people who have not been exposed to the suspected health risk but otherwise are as similar as possible to the group being studied. For example, if people in the experimental group smoke, people in the control group should be smokers too. Other important points of similarity might be age, sex, weight, race, or job (unless one of these conditions is the one being studied). If the suspected health risk really poses a danger, significantly more people in the experimental group should be ill than in the control group. However, in many of the initial studies of Love Canal residents, controls were not carefully chosen. In some cases, data from older studies were used instead of data from modern controls.

Another problem with some early studies was that the test groups were not selected randomly. In order for results to be accurate, those choosing the test group should have no previous knowledge of the possible candidates for the group. However, Love Canal residents did the first study themselves, interviewing fellow residents. Since they knew what they were looking for, their choice of whom to ask for medical histories could not be random. This was not the only study involving non-random selection of the test group. In an initial study on genetic damage, scientists selected residents known to have health problems instead of randomly selecting residents.

When these studies were reviewed by other scientists, these problems were discovered and tests were redone properly. However, before the new studies could be reviewed, the questionable results were leaked to the press, who treated them as conclusive. This frightened residents even more: the tests said that there were higher rates of illness and health problems in the Love Canal area. Thus, the residents continued to claim the chemicals were harming them and accused the government of hiding the truth.

Based on the questionable preliminary findings and pressure from frightened residents, the government authorized the evacuation and relocation—at government expense—of any family who wanted to move. Some residents chose to stay, never believing the chemicals posed a significant danger.

The entire site was cleaned and the dump was resealed. The school and homes closest to the canal were demolished and buried. The Hooker Chemical Company paid for the cleanup and was held solely responsible for the incident despite their warnings to the city and a ruling that said their disposal methods at Love Canal were acceptable within the rules for the 1940s and 1950s. Neither the board nor the city was blamed for the leaks, which were partially caused by excavations they authorized for the school and houses.

After the initial uproar, properly prepared and conducted studies were performed. These showed that in fact incidences of birth defects, miscarriages, cancer, and illnesses among the Love Canal residents were not higher than normal. The studies also could show no relationship between exposure to the chemicals and health problems residents reported. However, press coverage of these results was not nearly as extensive as coverage of the initial, faulty results which helped spread unnecessary panic and worry.

Today, the Love Canal is once again a neighborhood. The homes farther from the canal have been deemed habitable and are being sold. People grow gardens, play in their yards, and drink their water. Is there a danger from the chemicals? They don't think so. In fact, many of them feel safer, since the area is regularly monitored for leakage.

Were some of the original residents sick? Yes. Were chemicals leaking from the canal? Yes. Were the chemicals responsible for residents' health problems? No relationship has yet been found, but that does not mean that a relationship doesn't exist. Later tests may someday be able to prove conclusively that the chemicals were—or were not—to blame.

Think About It

1. Have you heard about Love Canal before? Did anything you read in this article surprise you?

2. What are the characteristics of a good control group? Why is a good control group necessary in research on health risks?

3. Why is it important to select a test group randomly?

Lesson 4:
Source Reduction

Students investigate the concept of waste reduction through two laboratory activities. The first, "**One Liter—To Go**," is a comparative packaging activity in which students calculate and compare the masses of different types of containers with a liquid capacity of 1 L. In the second activity, "**Wrap It Up**," students study the way chewing gum is wrapped. Students are asked to consider the importance of packaging in source reduction.

This lesson's *Garbage Gazettes* explore two problems of packaging. "**War of the Packing Fillers**" weighs the pros and cons of polystyrene versus starch loosefill packaging "peanuts." "**Enjoying Fresh Milk...In the Desert?**" traces the history of food preservation and explains the pros and cons of aseptic packaging for food items.

● Teacher Background on Source Reduction

A logical component of any effective integrated solid-waste management program is a strategy for reducing the amount of refuse entering the waste stream in the first place. "Source reduction" includes any action that reduces the volume or toxicity of solid waste prior to recycling or disposal.

Approximately 35% of municipal solid waste is packaging, so packaging is an important area to target for source reduction. In the United States, 5.6% of all steel, 50% of all paper, 65–70% of all glass, 25–30% of all aluminum, and 23.5% of all plastics produced are used for packaging. One out of every $10 spent on food in America goes into its packaging (Denison). Thus, American consumers pay 10 cents of every product dollar for something they usually discard. The following table, Amounts of Packaging in Municipal Solid Waste by Weight, shows different types of packaging and their percentage of total packaging waste in 1995. (EPA)

Amounts of Packaging in Municipal Solid Waste by Weight, 1995	
Packaging Type	Percent of Total Packaging in MSW Waste Stream
paper and paperboard	52.2%
glass	15.8%
plastic	10.6%
metals	6.6%
other	14.8%

Although at first it may seem desirable just to eliminate packaging altogether, packaging serves many useful purposes. Packaging can promote a product, protect its contents, discourage theft, divide a product into easy-to-use portions, and maintain cleanliness and safety. Food packaging can reduce waste by preventing spoilage. Packaging designers must consider these and additional factors, including the cost of materials, protection and transportation of the product, consumer convenience, product advertisement, and shelf appearance.

In order to analyze the costs and benefits of a packaging system, we must take into account all aspects of production, use, and disposal. First we must consider the commercial and environmental costs of the raw materials. Then we must consider the energy and other costs involved in manufacturing the packaging materials. For example, cutting, folding, shrink wrapping, and printing all require equipment, energy, and labor. Similar costs are associated with filling, sealing, and distributing the packages and their contents. These costs affect the design of the package. The nature of the product and the way it is most likely to be used must also be considered. For example, while single-serving packages require more material, they may also prevent spoilage and contamination, especially if the product will be consumed slowly (as in the case of one person taking a week to chew a package of gum). In other cases, packaging may make the product easier to use. Finally, we must consider the disposal of the package when it has served its purpose.

In order to fully evaluate the environmental impact of packaging, it is important to consider a package's entire life cycle as outlined above. The major packaging materials differ in their use of raw materials. Only wood- and paper-based packaging use a renewable resource. Glass and steel packaging are made not from renewable resources but from naturally occurring raw materials that are in plentiful supply. Tin and chromium for coating steel are both relatively scarce, but because the amounts required are comparatively small, supply does not appear likely to be a problem in the near future. Aluminum is plentiful, but ore-grade resources of this metal are much less common. Plastics are based on petroleum and natural gas, which are also valuable as energy resources. For all these materials, recycling of packaging can extend the life of the raw material base by displacing virgin materials with recycled ones. However, some raw materials will continue to be needed as demand increases and because the recycling rates for all of these materials are less than 100%.

Production costs vary depending on the type of packaging material. Glass is energy-intensive to manufacture. However, refillable glass bottles require the least energy to recycle of all single-service beverage containers. Recycling non-refillable glass bottles requires more energy than recycling aluminum cans. Producing aluminum packaging requires large amounts of energy in refining the metal from ore, but using recycled aluminum requires much less energy. Petroleum and natural gas are needed as both ingredients and energy sources in plastics manufacturing. This energy can be partially recovered through incineration. Plastics generally require less energy to produce than equivalent paper products. Almost all paper recycling processes use significantly less energy than manufacturing paper from all raw materials.

The distribution of packaging materials, along with the product, requires extra energy. Glass packaging, for example, may contribute significantly to transportation energy costs due to its great mass. Whether the material is landfilled or recycled, further energy is required to transport it after it is discarded.

● Teacher Notes for "One Liter—To Go"

In this activity, your students will investigate the different packaging choices for one particular product. They will determine and compare the masses of different kinds of packaging needed for 1 L of liquid product.

Group Size .... 3–4 students
Getting Ready .... 5–10 minutes
Time Required .... Getting Ready: 5–10 minutes
                                            Procedure: 40 minutes
                                            Discussion: 20 minutes

Safety and Disposal

No special safety or disposal procedures are required.

Materials

For Getting Ready
Per class
- food color
- water
- easy-pouring container, such as a 2-L plastic bottle or 1-gallon plastic milk jug
- can opener
- masking tape

For the Procedure
Per class
- volume measuring devices such as 250-ml graduated cylinders or metric measuring cups
- balances

Per group
- clean, empty container for drinkable liquids, such as one of the following.
  - 1-L (33.8-ounce) PETE bottle with lid (#1 recycling code)
  - 0.5-L (16.9-ounce) PETE bottle with lid (#1 recycling code)
  - 1-L (33.8-ounce) glass bottle with lid
  - 473-mL (16-ounce) glass bottle with lid
  - 1-L (33.8-ounce) gable-top cardboard carton (such as a milk carton)
  - 240-mL (8-ounce) gable-top cardboard carton

- 355-mL (12-ounce) aluminum can
- 1-L (33.8-ounce) bimetal can (such as a tomato or pineapple juice can)
- 177-mL (6-ounce) bimetal can
- 250-mL (8.5-ounce) aseptic carton (such as a juice box)
  - 1 L colored water (see Getting Ready) in a separate container

For the Extensions

Per class

- aseptic drink box
- bowl of water
- hand mixer

Getting Ready

Ask your students to bring in as many different kinds of packages that have held drinkable liquids as they can find. The containers must be empty and clean. Try to collect a representative sample of the different types and sizes of containers listed in Materials.

Rinse all containers carefully and set them aside to dry before use. Remove the section of each label that shows each container's volume. Make a large enough opening in the bimetal containers so that students can easily pour liquids into and out of them and to ensure the container has been cleaned out thoroughly. Cover any sharp edges with masking tape. Keep the lid of each container with that container.

For each group, measure 1 L water into an easy-pouring container, such as a 2-L plastic bottle or 1-gallon plastic milk jug. Add one or two drops of food color to each group's water.

Opening Strategy

1. Display a representative sample of each kind of container the students brought in. Ask students how the containers are different, and list some differences on the board. Ask what effect these different factors would have on transporting and disposing of the containers. Focus on the mass and volume of discarded materials with regard to the issue of source reduction. Ask what effect opacity might have on the shelf life of products, such as some foods or medicines. Also, ask students to think about product tampering safeguards.

2. Discuss how students might go about comparing the mass of various types of packaging. Ask why simply comparing the masses of different containers does not provide useful information. Lead students to conclude that they need to compare the mass per unit of volume.

3. Challenge each group to determine the mass of packaging for 1 L of a liquid product using one of the types of packaging displayed. (For the purposes of this activity, they may calculate fractions of a container; they do not have to round up to the next whole container.) Give each group a different container from the selection, 1 L colored water, and the Student Instructions, and have them proceed.

Depending on the grade level and experience of the students, you may wish to help them with the calculations in steps 4 and 5 of the Student Instructions. The number of containers they need to hold 1 L is calculated by dividing 1 L by the volume of their container in liters. The mass of containers needed to hold 1 L of the product is calculated by multiplying the number of containers needed by the mass of one empty container.

Extensions

1. Have students take their findings home to share with their families. Encourage them to accompany a parent or guardian on a trip to the grocery store to help make purchasing decisions based on what they learned about packaging.

2. Show students how drink boxes can be recycled. Rip one aseptic drink box into several smaller pieces. Allow the pieces to sit in a bowl of water overnight. The next day, use a hand mixer to pulp the contents for several minutes. Have students observe the pulp. How many different materials are apparent? What are these materials? Which of these materials can be recycled? Does it seem practical to recycle these boxes?

3. Investigate how many different types of packaging could be used for a given product.

Cross-Curricular Integration

Business, marketing, and economics:
• Ask the class to define what packaging characteristics they consider to be environmentally friendly. Have students design a label for a new fruit drink in an environmentally friendly container and incorporate the packaging information into their advertising campaign.

Language arts:

- Ask students to write letters to companies who they think use excessive packaging in their products telling them of their findings and suggesting alternatives.

Social studies:

- Discuss the history of packaging and how it has changed over time. Take into account amounts of packaging, materials available, and transportation methods.

● Student Instructions for "One Liter—To Go"

In this activity you will consider how similar products can be packaged in many different ways. The investigation focuses on beverage packaging as a source of solid waste and the role consumer choice can play in source reduction.

Safety and Disposal

No special safety or disposal procedures are required.

Procedure

1. The 1 L of colored water your group received represents a new product that needs packaging. Your group is part of the package development team. Begin by naming the product and recording its name in the Data Recording section of this handout. Record the type of container both on the second line of the Data Recording section and in the Packaging Mass Comparison table next to your group number. Circle your group number.

2. Determine the mass in grams of your container and record it.
    Don't forget that lids and can tops are part of the total packaging and should be weighed with the containers.

3. Use the colored water and appropriate measuring devices to determine the volume of product your container holds (in liters) and record the result.

4. Calculate how many of your size of container would be needed to hold 1 L of the product. Show your calculation and record the result both on the fifth line of the Data Recording section and in the Packaging Mass Comparison table next to your group number.

5. Calculate the total mass of the containers needed to hold 1 L of the product. Show your calculation and result. Also record the result in the third column of the Packaging Mass Comparison table in the row for your group number.

6. Obtain data from the other groups about their containers and record it in the Packaging Mass Comparison table.

7. Answer questions a–f in the Analysis Questions section of this handout.

Data Recording

_____ Product name

_____ Type of container (size and material)

_____ Mass of one container (in grams)

_____ Volume of product that one container holds (in liters)

_____ Number of containers of the size you have that are needed to hold 1 L of product
Show your calculation:

_____ Total mass of containers needed to hold 1 L of product
Show your calculation:

Table: Packaging Mass Comparison

Group number	Type of container used (volume and material)	Total mass of containers needed to hold 1 L of product
1		
2		
3		
4		
5		
6		
7		
8		
9		

Analysis Questions

a. If mass were the only consideration, which type of packaging would you recommend as best? Why?

_____

_____

_____

_____

b. Another factor to consider is the amount of space the discarded beverage container takes up in a garbage truck or landfill. Which type would you recommend if this were the only consideration?

_____

_____

c. What other factors (such as convenience, price, portability, breakability, and availability of refrigeration) might make a particular kind of packaging desirable? When might you choose differently? For example, how would beverages you purchased for a sack lunch be packaged? For a class party? For a camping trip?

_____

_____

_____

_____

_____

_____

_____

d. Consider the differences between single-serving containers and larger containers. If you could serve eight people with either 8 cans or one 2-L bottle of soft drink, which would you choose? Which would generate more packaging waste?

_____

_____

e. Discuss other elements of the different types of packaging, such as how well each protects the product and whether the product could react with the packaging in an undesirable way. (For example, the liquids could not be packaged in paper.)

_____

_____

_____

_____

_____

_____

f. Is the packaging you recommended as best in question "a" also the best when these other factors are considered? Why or why not?

_____

_____

_____

● Teacher Notes for "Wrap It Up"

Students will determine the amount of packaging used for a familiar product.

Group Size .... 2–4 students
Time Required .... Getting Ready: 10 minutes
                                                                  Procedure: 45 minutes

Safety and Disposal

No foodstuff should be consumed in any laboratory situation, so allowing students to chew the gum used during the activity may mislead them about proper laboratory procedure. Therefore, if gum chewing is allowed in the classroom, you may wish to purchase additional gum for students to chew after the lab has been completed and cleaned up.

No special disposal procedures are required.

Materials

For the Procedure
Per class

- 6–7 different brands of chewing gum with unique packaging

  Get enough packages so that each group can have one. Be sure to record the price of each brand of gum at the time of purchase.

- electronic or triple-beam balances

  If an electronic or triple-beam balance is not available, as an alternative use brands of gum whose mass is the same according to the package and then have the students compare the masses of the packaging using a two-pan (double-pan) balance.

For the Extensions
Per class
- different types of cereal or candy
- electronic or triple-beam balance

Opening Strategy

1. Ask the students if any of them chew gum. If so, what factors do they use to select the brand they buy? Have they ever selected gum just because of the packaging? Was the packaging in some way useful enough to keep, or did they throw it out after removing the gum?

1b What is source reduction?

2. Using less packaging is one of the simplest ways to reduce the amount of discarded material entering the waste stream. One way to evaluate packaging for this purpose is to calculate how much of a packaged product's mass is packaging. Tell students they will conduct a small-scale study in source reduction by investigating the mass of gum packaging and its percentage of the total mass of the packaged product.

3. Show the students the different brands of chewing gum for use in the Procedure and list the names and prices on the board.

4. Hand out the Student Instructions and packages of gum to the groups and have them follow the Procedure.

Extensions

1. Perform this activity with different types of cereal or candy.

2. Have students design a totally new way of packaging gum with the least amount of packaging that would keep the product clean. Would selling gum in bulk be an option? Have the students seen gumball dispensing machines? How about reusable packaging? Have students try to sell their products to classmates. What strategies work best to sell a product?

good idea

Cross-Curricular Integration

Home, safety, and career:

- Use this activity to introduce students to the topic of money management. If they had a budget for their gum allowance, how could they get the most gum for their money? If they compared different sizes of gum packages, was the gum less expensive per gram when purchased in large packages? What is the most economical way of buying food for a family? Is this way necessarily the best way for a single person?

- Students could use the World Wide Web to investigate awards for good and bad packaging that are issued by various organizations.

- Discuss the proper disposal of chewed gum and the importance of protecting both human health and property. Incorporate the idea of reusing the wrapping to dispose of gum.

good idea

Language arts:

• Have students write to the manufacturers of the brands of gum that used the most packaging. Students should list their findings and make suggestions for more efficient packaging.

Mathematics:

• Have students create and record results of a survey asking other students what brand of gum is their favorite and why.

● Student Instructions for "Wrap It Up"

In this activity, you will consider the amount of packaging used to wrap a familiar product.

Safety and Disposal

No foodstuff should be consumed in any laboratory situation. Do not chew the gum used in this activity.

No special disposal procedures are required.

Procedure

1. Record the different brands of gum and their prices in the appropriate columns on the Data Recording table in the Data Recording section of this handout. Remember to match each brand with the number of the group that is working with it.

2. Find the total mass (gum plus packaging) of your group's package of gum. Record the mass in grams in the fourth column of the Data Recording table.

3. Record the gum's mass (printed on the package) on the Data Recording table.

4. Calculate the mass of the packaging using the total product mass found in step 2 and the printed gum mass found in step 3. Record your result on the Data Recording table.

5. Unwrap all the pieces of gum in your package. Put all of the pieces of gum in one pile and all of the packaging in another pile.

6. Measure the mass of the unwrapped gum and record it on the Data Recording table.

7. Find and record the mass of the packaging alone.

8. Calculate the packaging percentage using the measured mass of the packaging and the measured total mass of the gum plus packaging. Record your results in the Data Recording table. Graph your result by writing the name of your brand of gum beside your group's number on the Percentage of Packaging in Total Product Mass graph and coloring in the corresponding row to the appropriate percentage.

9. Use the cost of the gum and the measured mass of the gum in the package to determine the cost per gram of gum and record it in the last column of the Data Recording table.

10. Obtain and record the data for different packages of gum from all of the other groups. Record this information on the table and add the package percentages to the graph.

11. Discuss questions a–n in the Analysis Questions section of this handout as a group and record your answers.

Data Recording

Table: Data Recording

Group number	Brand of gum	Product cost	Total product mass	Printed gum mass	Calculated gum mass	Actual gum mass	Actual packaging mass	Packaging percentage of total mass	Cost per gram of gum
1									
2									
3									
4									
5									
6									
7									

Graph: Percentage of Packaging in Total Product Mass

Group number	Brand of gum	10%	20%	30%	40%	50%	60%	70%	80%	90%	100%
1											
2											
3											
4											
5											
6											
7											

Analysis Questions

_____ a. Is the measured mass of the gum the same as the mass printed on the label?

_____ b. Do the measured masses of the gum and the packaging add up to the measured total mass?

_____ c. Which brand had the lowest percentage, by mass, of packaging?

_____ d. Which brand had the highest percentage, by mass, of packaging?

_____ e. If every student in the class chewed one package of the gum with the highest packaging percentage, how much packaging would the class be throwing away?

_____ f. If each student in the class chewed one package of the brand of gum from question e per day for a whole school week (five days), how much packaging would the class be throwing away?

_____ g. How much packaging would be thrown away in a month?

_____ h. How much packaging would be thrown away in a year?

i. Describe how the packaging differed between the brand with the highest percentage, by mass, of packaging and that with the lowest.

_____

_____

j. What reasons might the manufacturers have for packaging their products the way they do?

_____

_____

k. How could the manufacturer reduce the amount of packaging and still accomplish the same thing?

_____

_____

l. Can any of the packaging be reused or recycled? How?

_____

_____

m. Compare the price of each brand of gum to the amount of usable product. Which brand was least expensive per gram? Which was the most expensive? Was the most expensive gum also the one with the most packaging?

_____

_____

_____

n. Which brand is the most popular with your group? Why is this gum the favorite?

_____

_____

_____

The Garbage Gazette

January 6   Local Edition   Vol. 1, Issue 6

War of the Packing Fillers

If you've ever received a box in the mail, chances are it contained more than just what you ordered or what a relative or friend sent you. It probably contained some sort of "filler" to protect the contents—"peanuts," bubble wrap, shredded paper, or molded foamed polystyrene (popularly known by the trade name Styrofoam™). Once you've retrieved the goodies, what do you do with the filler? What *should* you do with it?

Questions like these have sparked debate among industry, environmental organizations, and the general public. Everyone agrees that filler needs to be sturdy and protective, but each group has its own ideas about what materials should be used to make it. Companies must consider cost, material source, environmental impact, customer opinion, and other factors before deciding on a fill to use or make.

At the forefront of this debate is the most popular form of filler known as peanuts. Peanuts fall into a category of filler called "loosefill," which means that the filler consists of small pieces of material that are able to move around in the box. Molded polystyrene, on the other hand, forms an unmoving framework that holds the contents of the box (such as stereos or small appliances) in place. In 1990, an organic, cornstarch-based alternative to foamed polystyrene loosefill was developed in response to public opposition to

Can you tell the difference? Which of these peanuts are polystyrene and which are cornstarch?

foamed polystyrene products. Is the starch loosefill, known by its trade name Eco-Foam®, better than polystyrene? Before shipping companies can make this decision, they need to look more closely at each type of loosefill.

Foamed polystyrene loosefill. Styrene is an organic hydrocarbon produced from ethylbenzene, a compound derived from petroleum.

Foamed polystyrene is not water-absorbent and makes an excellent electrical insulator. These characteristics make polystyrene an effective packing filler.

Starch loosefill. In the late 1980s, Paul Altieri and Norman Lacourse, employees of National

Starch and Chemical Company, were experimenting with processes that would increase cereal tolerances to milk. Most of their experiments included sending cornstarch-based cereals through an extruder. During extrusion, heat and pressure cooked the starch, which turned into a gel. As the gel leaves the extruder, the heat and pressure are released, causing the starch to foam up. Occasional batches of "cereal" from the extruder were spongy and flexible like polystyrene rather than hard and brittle like cereal.

Instead of tossing aside the cereal misfits, National Starch and Chemical began exploring this extruded starch project to see if it could become a viable alternative to polystyrene. Their research has developed a special corn hybrid that produces the most durable, most protective starch loosefill, which they have patented as Eco-Foam. Then, they began working with American Excelsior Company to refine the Eco-Foam to be used in the packing industry.

Which would you choose? Studies have shown that both forms of loosefill are about equally as durable and protective. Also, because both have excellent cushioning ability, packagers can use relatively small volumes of either loosefill. Both types of loosefill flow through packing machinery easily and efficiently. If both types perform equally as well, how can companies decide which to choose? What other factors are important for the decision? Read the chart below, then answer the questions in the next column.

Think About It

1. Look at the sources the information in the chart below came from. Might these sources be biased to present the information about the different loosefills in a certain way?

2. If you were sending a package and had to choose a loosefill, would you use polystyrene fill or starch fill? Why did you make this choice?

	Foamed Polystyrene Loosefill (a.k.a. Styrofoam™)	Starch Loosefill (a.k.a. ECO-FOAM®)
Ingredients	Polystyrene and a blowing agent (usually pentane or CO_2)	Cornstarch, polyvinyl alcohol, and water
Biodegradable?	No. Polystyrene won't biodegrade.	Yes, unless it is placed in a landfill. Nothing biodegrades in landfills, not even products made from plants, because landfills do not get enough air and water.
Reusable?	Yes, as packing foam, a soil conditioner, or a clean-burning fuel in waste-to-energy incinerators	Yes, it can be used as packing foam or it can be placed in a compost pile or on a lawn or garden (will dissolve after it rains)
Recyclable?	Yes. Companies that use it support its recycling.	No information was found on starch loosefill recycling.
Renewable resource?	No. It is derived from petroleum and natural gas.	Yes. It is derived from corn, a renewable crop.
Cost	Costs about half as much as starch loosefill	Costs about twice as much as polystyrene loosefill
Misconceptions and Facts	Many people believe foamed polystyrene is still made with chlorofluorocarbons (CFCs), which deplete the ozone layer. This is not true. CFCs have not been used in polystyrene production since 1990. When they were used, polystyrene production accounted for only 2–3% of the CFCs used in America. People also believe that polystyrene, especially foamed polystyrene loosefill, takes up a lot of space in landfills because it is usually thrown away, not reused or recycled. This is not true. Polystyrene takes up only 0.9% of landfill space, with ¼ of that being loosefill.	Many people believe that plant-based materials will biodegrade in landfills, so they might throw away starch loosefill. Shippers worry about what will happen to the water-soluble loosefill if the package gets wet or is shipped in humid conditions. However, because starch loosefill must soak in water before it will dissolve, other items inside a package sitting that long in water will probably also be ruined anyway.
Source of information	Plastic Loose-Fill Producers' Council, Polystyrene Packaging Council	National Starch and Chemical Company (inventor), American Excelsior Company (manufacturer)

The Garbage Gazette

January 27   Local Edition   Vol. 1, Issue 7

Enjoying Fresh Milk...In the Desert?

Just as you eat food to give you the nutrients you need to live and grow, other living things are doing the same thing. Some of these organisms are so tiny they can be seen only with a microscope, and we call them microorganisms. But if you can't see microorganisms with your naked eye, how do you know they're around?

Various clues indicate the presence of microorganisms. The blue-green fuzz on your bread tells you that mold spores have set up camp. A sour smell is a sign that a colony of bacteria is inhabiting your milk. When microorganisms leave these signs, we say that the food is "spoiled."

Food spoilage falls into different categories. Cheese and yogurt are foods that are deliberately "spoiled." Harmless bacteria cultures are added to milk, and as these cultures grow, they turn milk into cheese or yogurt. Different bacteria produce different flavors and textures. Mildly spoiled foods, like moldy bread or slightly soured milk, probably wouldn't kill you if you consumed them accidentally, but they would make you sick to your stomach. However, some spoiled foods can make you very sick or even kill you. For example, foods contaminated with *Clostridium botulinum* contain a lethal poison that the bacteria produces in an airless environment, such as a sealed can or jar.

In order to prevent or slow the growth of microorganisms, food must be preserved. Food preservation techniques have been around as long

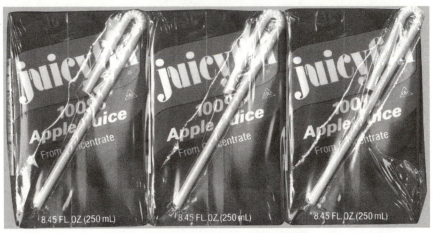

Aseptic packages like these drink boxes were called the top food science innovation since 1939 by the Institute of Food Technologists.

as humans have. Early people dried their food in the sun to make it last longer. They didn't know why that worked, but now we do: microorganisms, like people, need water to grow and survive. Drying removes all or most of the water, slowing microorganism growth. The horsemen of Ghengis Khan used a dehydration technique to carry milk with them as they set out to conquer Asia in the early 13th century. They carried two bags, one containing dried milk powder and the other containing water. Whenever they needed milk, they mixed the powder with water.

Another early food preservation method was "curing," which involved soaking food in a salt, sugar, or acidic solution. These solutions inhibit microorganism growth, because salt, sugar, and acids dehydrate food. Food can also be cured by smoking, a process in which food is placed in a special "house" over a slow-burning fire.

Chemicals in the smoke are absorbed by the food and slow the growth of microorganisms.

Although drying and curing are effective in slowing the spoilage rate of food, they also significantly change the foods. Fruits take on a new texture after drying, and pickles cured in a vinegar solution taste little like the cucumbers they were originally. Other methods, like chemical additives, refrigeration, and pasteurization, have little effect on the taste and texture of the food, but they delay spoilage for only a short time.

In the early 1800s, a French candy maker named Nicolas Appert developed a long-term preservation method that had little or no effect on the taste and texture of food. He cooked the food and placed it in sealed jars, which was an effective short-term technique. Appert went one step further and boiled the sealed jars in water, which destroyed more microorganisms than other methods and delayed spoilage longer.

Appert's technique is now known as the aseptic packaging technique. When foods are aseptically packaged today, they are first subjected to an ultra-high temperature (UHT) treatment that destroys microorganisms while retaining most of the flavor and nutritional value. The food is then packaged in a sterile container made of layers of paper, plastic, and aluminum foil. This container keeps out microorganisms, water, ultraviolet light, and oxygen and other atmospheric gases, which can cause spoilage or break down nutrients.

Food prepared using the aseptic technique does not need refrigeration or chemical preservatives. Milk preserved this way can stay fresh for six months or longer without refrigeration, even in hot climates. Aseptically packaged milk was used by troops serving in the Persian Gulf War in 1991.

Although aseptic packaging was considered the #1 food science innovation since 1939 by the Institute of Food Technologists in 1990, some people protest its use. These people are mainly concerned about the recyclability of the packages.

Although aseptic packages have always been recyclable, recycling programs are not necessarily feasible for all communities. Aseptic packages are a mixed-material package and must be specially processed to be recycled. However, not all communities have access to mixed-material recycling facilities. This led people to believe that aseptic packages could not be recycled. However, aseptic packages can be recycled in two ways. In "hydrapulping," the boxes are stirred in water until the plastic, paper, and foil layers separate. The paper is recovered for paper-milling,

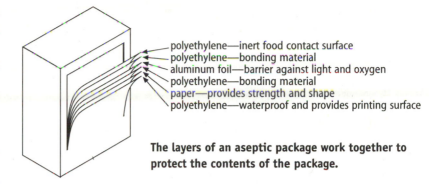

polyethylene—inert food contact surface
polyethylene—bonding material
aluminum foil—barrier against light and oxygen
polyethylene—bonding material
paper—provides strength and shape
polyethylene—waterproof and provides printing surface

The layers of an aseptic package work together to protect the contents of the package.

and the foil/plastic residue can be added to mixed-plastic resources. In the second option, aseptic packages are mixed with plastics of all kinds to make a new plastic mixture. The paper and aluminum of a drink box are such a small percentage of the overall mixture that they do not need to be removed.

To help reduce opposition to the drink box, the Aseptic Packaging Council (APC) has set up drink-box recycling programs in many schools and residential areas.

Besides availability of recycling programs, another objection to drink-box recycling is that it does not operate in a closed loop—that is, drink boxes cannot be recycled into new drink boxes. Thus, new raw materials are always required to make more drink boxes.

While critics of aseptic packages usually concentrate on the recycling aspect, aseptic packaging does have environmental benefits. Aseptic packages have the lowest ratio, by weight, of packaging to contents, which means that fewer materials are used to make these packages than are needed to make any other beverage container. A drink box is about 96% beverage and 4% packaging, aluminum cans 95% beverage and 5% packaging, and glass bottles 63% beverage and 37% packaging.

Aseptic packaging also uses less energy in transport and storage than other packaging. One truck can transport 1.5 million empty drink boxes, but it takes 14 trucks to transport the same number of glass bottles. In addition, because aseptic packages don't require refrigeration, transporting and storing them requires less energy and reduces refrigerant emissions.

Despite opposition, drink boxes are still a popular beverage container option because they are strong and durable and preserve beverages for longer periods of time. In 1992, more than 8,000,000 drink boxes were used every day. Manufacturers are also beginning to use aseptic packages for vegetables, liquid eggs, soup, and pancake syrup. And as technology advances, the industry may improve the recycling technique enough to win over the opposition, making the drink box the container of the future.

Think About It

1. *What are some preservation techniques people have used to keep their food fresh? What are the pros and cons of each method?*

2. *What is an aseptic package made of? How do these materials work to protect the contents of the box?*

3. *Do you think it is more important to keep food safe or to create easily recyclable packaging? Why?*

Lesson 5:
Reusing Materials

The idea of reusing materials that would otherwise be thrown away is addressed through two activities that involve making products from items often thrown away. In the first activity, "**Treasures from Trash,**" students make original inventions or artwork from objects that would otherwise be discarded. In the second activity, "**Shrinking Crafts,**" students design and make their own shrinking craft items from discarded polystyrene.

Students learn about people who reuse materials on a grander scale in *The Garbage Gazette* "**Earthships Take Off,**" a description of houses that are built using old tires, pop cans, glass bottles, and other already-used materials. *The Garbage Gazette* "**Don't Dismiss Disposables**" looks at the debate about plates, cups, "to-go" packaging, and other food-service items in restaurants.

● Teacher Background on Reusing Materials

● ● ● ● ● ● ● ● ● ● ● We can easily fall into the pattern of believing that newer is always better. Unfortunately, many perfectly good products are routed into the waste stream every day to make way for new products. These discarded items often still have value; they just need to be repaired or repainted or to have a new use found for them.

By some estimates, up to 10% or more of the solid waste stream is made up of items that could be reused fairly easily. Obvious examples include clothing, books, toys, and furniture. Many recycling centers have some type of swap shop to store and distribute such items. Sometimes local social service agencies provide collection points at recycling facilities. Reusing products saves the costs of landfilling those products, and the individuals taking the items are getting something they value. When reuse is possible, it is an even more beneficial waste reduction strategy than recycling, which requires energy for trucks to collect items, machines to sort them, and machines to make them into new products.

Antique stores, flea markets, thrift shops, and garage sales are all good sources for used products. Furniture, clothing, decorative items, and other treasures may be stored in neighbors' basements or attics just because they are no longer useful to that particular owner. Buying previously owned items can be good for the environment as well as for the budget.

People can also prolong product life in their own homes. The useful life of many existing products can be extended by a little bit of repair or touch-up work. Typically discarded items can be reused as parts of homemade crafts or art. To many people, handmade products such as the ones students make in the following two activities have a value that far exceeds the value of the materials used.

● Student Background on Reusing Materials

When people today want a machine that washes dishes automatically, they can go to an appliance store and buy one. Prior to 1886, people didn't have this luxury, because the dishwasher hadn't been invented yet.

Josephine Cochrane was an Illinois woman who enjoyed hosting dinner parties but didn't like the cleanup process. She wanted a way to clean her dishes quickly and without breakage. Since there was no such process, she invented one: using old, discarded materials from around her house, she created the first dishwasher. She fashioned wire into custom-made compartments for cups, saucers, and plates, then set these compartments into a wheel in the bottom of a copper boiler. While hot soapy water sprayed up through the boiler, an old motor turned the wheel. Cochrane debuted her invention at the 1893 World's Fair, where it won the highest award.

Instead of throwing away the old wire, wheel, and boiler, Cochrane made a dishwasher. She reused items that otherwise would have ended up in the solid waste stream, thus saving the cost of transporting, landfilling, and/or recycling the materials. Cochrane isn't the only inventor who has made use of discarded materials: so-called junk has been used for some amazing inventions, such as the following:

- Early peoples used animal spines and thorns as hairpins. Dried fish backbones became combs.

- The plane the Wright Brothers flew at Kitty Hawk was made in part with used bicycle parts.

- Earl Tupper bought a mass of smelly polyethylene waste from DuPont, who was going to throw it away. Tupper refined the waste to create Tupperware®, a line of food storage containers.

- Edward Pauls built the first NordicTrack® from parts he mined in his neighborhood junkyard.

- An 18th-century Dutchman sewed wooden spools to his shoes and created the first roller skates. In the 1980s, Scott and Brennan Olsen attached wheels to a pair of old ski boots to create the Rollerblade® brand of in-line skates.

- James Wright was working on a new type of synthetic rubber when he came up with a rubber-like compound that stretched longer and rebounded higher than other synthetic rubbers. However, these characteristics did not give it enough advantage over other synthetic rubbers. Instead of throwing away the substance, Wright sent it to engineers to see if they could find a use for it. At a party where this "nutty putty" was being demonstrated, Ruth Fallgetter and Peter Hodgson realized this putty would make a great toy, and so Silly Putty® was born.

- When 19th-century Americans were finished reading catalogs, newspapers, and advertisement fliers, they used them as toilet paper.

- Workers at Robert Cheeseborough's kerosene plant were plagued by a pasty, paraffin-like residue, a by-product of the petroleum processing process, that gummed machinery into inactivity. However, the workers learned that if they rubbed it on a wound or burn, healing was accelerated. The substance eventually was marketed as "Vaseline®."

Inventors are not the only people who turn junk into valuable creations. Artists also transform already-used materials into artwork—sometimes for the sake of beauty and sometimes to send a message. Mindy Lehrman Cameron is such an artist. She used large amounts of garbage to create an exhibit on solid-waste management at the Connecticut Resources Recovery Authority visitors' center in Hartford. The exhibit includes a "Temple of Trash" with columns made of used tires; interactive displays constructed out of such things as plastic bottles, recycled carpet, brooms, and shovels; and placards mounted on aluminum trash cans. To create and mount the various displays, Lehrman Cameron worked hard to use only environmentally sound components.

In Detroit, artist Tyree Guyton uses garbage to combat urban decay in the neighborhood where he grew up. Guyton's "Heidelberg Project" takes found objects, such as used shoes, broken dolls, fire hydrants, and old toilets, to decorate abandoned houses and other neighborhood landmarks. With the help of many people, including his 92-year-old grandfather, Guyton paints empty buildings, old oil drums, fireplugs, and signs with colorful stripes, squares, and polka dots. To trees and porches he mounts other already-used materials: stuffed animals, old crutches, chairs without legs, and inner tubes. With his painting and his sculptures he hopes to bring attention to urban decay and to make his own neighborhood safer and more beautiful.

The Museum of International Folk Art in Santa Fe, New Mexico, sponsors an exhibit entitled "Recycled, Re-Seen: Folk Art from the Global Scrap Heap." Specimens from this fascinating exhibit can be viewed at http://www.state.nm.us/MOIFAOnLine/RecycledReSeen/RRtempindex.html.

Although many of the objects in the exhibit were created more out of necessity than recreation—people in other countries mine dumps to find materials they can transform into usable objects and sell for a living—they all reuse discarded objects. Examples include a ham can violin, dustpans made from license plates, buckets and water jugs made from old tires, and briefcases made from tin cans. The exhibit also features the steel drum, which originated from discarded oil barrels.

The steel drum, which was invented in Trinidad, has an interesting history. During the late 19th century, the British government, which was in control of Trinidad at the time, outlawed all native drums, which were used in gang wars. Deprived of their traditional instruments, the people had to use whatever they could find as rhythmic instruments for their parades and music. In the late 1930s, someone discovered that a dented section of a barrel head produced a tone. A rich supply of such barrels was the large number of discarded 55-gallon oil drums from World War II naval bases on the island.

The British performance art group Yes/No People also reuses materials that have been discarded to make music. The group, who produces the hit show "Stomp," uses trash in their skits, including trash-can lids and brooms as the props for their choreography. To prepare for a show in a new location, they rummage through local garbage heaps to find new "instruments." By doing this local research, they can customize the show for its locale.

Reuse of discarded materials does not have to be reserved for creative projects, however. It can also be incorporated into daily life. The Burger family of Whitney Point, New York, throws away 3 pounds of trash a year, in sharp contrast to the American average of 4.3 pounds of trash per person per day. The Burger family secret: they reuse as much as possible, even going so far as to wash plastic straws and sandwich bags to use again. Sara Steck of Berkeley, California, turns cereal boxes and magazine pages into postcards and greeting cards. Her family also reuses grocery bags and takes cloth lunch bags instead of paper ones to work and school. The National Recycling Council suggests more reuse ideas, including using old paper as scrap paper before recycling, donating unwanted furniture and appliances instead of throwing them out, and washing and reusing plastic cutlery and aluminum foil.

● Teacher Notes for "Treasures from Trash"

Reuse is a desirable alternative to landfilling, incineration, and even recycling because it removes items from the waste stream, or at least postpones their entry. Also, it can decrease the need for raw materials and energy. In this activity, the students will observe a bird feeder made from already-used materials and then design their own inventions or art from common materials that are generally considered to be trash.

Group Size .... 1–4 students
Time Required .... Getting Ready: 20 minutes
Procedure: 60–80 minutes over 2 days

Safety and Disposal

Remind students to use caution when working with materials that have sharp edges. If necessary, provide students with chalkboard erasers to support containers from the inside, and supervise them closely as they poke holes through the containers. For younger students, holes should be punched by an adult. Consider safety when evaluating all of the inventions and artwork, and discuss safety with each student or group.

No special disposal procedures are required.

Materials

For Getting Ready
Per class
- at least 1 of the following:
  ○ plastic milk or water jug
  ○ cardboard gable-top milk or juice carton — *ie: Minute Maid cardboard juice container*
  ○ plastic 2-L soft-drink bottle
  ○ 2 aluminum pie plates
  ○ plastic cup
- 2–3 sticks, pencils, or dowels
- scissors
- wire cutters
- 0.5–1 m light wire
- hole punch

For the Procedure

Per group

- several sheets of paper
- previously used materials such as the following:
  - containers such as small boxes, plastic milk or water jugs, cardboard milk or juice cartons, plastic soft-drink bottles, or plastic tubs (margarine tubs)
  - aluminum pie plates
  - disposable cups
  - bottle caps or buttons
  - paper, paperboard, or cardboard
  - sticks
  - pencils
  - empty pens
  - twist-ties
- scissors
- ruler

Per class

- hammers
- pliers
- hole punches
- files
- sandpaper

Getting Ready

Ask students to bring in some of the large materials, such as plastic soft-drink bottles and milk or water jugs and cartons. Rinse all containers thoroughly and set them aside to dry before the activity.

Make a bird feeder using common household materials. Refer to the following figures for ideas or create your own design.

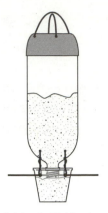

2-L bottle bird feeder

aluminum pie pan bird feeder

milk jug bird feeder

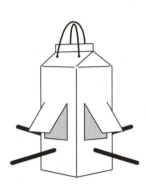

gable-top carton bird feeder

Opening Strategy

1. Tell your students to imagine they are working for a community wildlife conservation group. Alarmed by the declining numbers of songbirds in their neighborhood, the group wants to address this problem, as well as raise needed funds, by producing and selling bird feeders. They decided it would be most economical and environmentally sound to use materials that would otherwise go to the local landfill to make an easy-to-construct, unique, and functional bird feeder.

2. Lead the students in developing a list of factors they should consider in the design process, such as where the birds will perch, how much seed the feeder will hold, and how the feeder will be suspended.

3. Show the students the feeder you designed. Discuss the design features of this feeder in terms of the factors the class listed.

4. Have students read the Student Background (provided).

5. Discuss the various objects people have made from reused materials. Have students ever made anything from something that would otherwise have been thrown out?

6. Tell students that they will have a chance to design their own creations (inventions or art) using only already-used materials. They may not use anything new, including glue or tape. Tell students they may use the materials that have been collected, but they may also bring a few extra materials from home, as long as those materials were going to be discarded. Review students' designs and construction processes for safety and practicality before allowing them to make their creations. When students have completed their creations, have them share their creations with the class.

 Students may work either individually or in groups to design and make their creations. The construction phase can be done a day or two after the design phase to allow students to bring materials from home. For safety reasons, discourage students from digging through trash cans to find materials. Also, discourage students from bringing most of their materials from home; they may be tempted to bring materials that were not really going to be discarded, which defeats the purpose of the activity.

Extensions

1. Provide students with price lists for some of the materials they used to construct their creations. If they used materials other than those you provided, have them estimate the costs of these materials. Have them calculate the construction cost of their design, including the cost of labor. How might this compare to the cost of commercial objects similar to theirs?

2. Hold a "creation convention" to allow students to show off their artwork and inventions. Invite families and/or other classes to view the creations. Have students demonstrate their creations and answer questions about how they built them and the materials they used.

3. Have students make model planes from aluminum cans. Patterns for planes like the one on the first page of this lesson are available from B.C. Air Originals, P.O. Box 331, Colville, WA 99114, command@bcairoriginals.can. Visit B.C. Air Originals' World Wide Web site at http://www.bcairoriginals.com. Each series contains step-by-step instructions and templates drawn to scale. The difficulty level of this project is similar to that of making other model planes. The project requires sharp tools and should be done only by older students under adult supervision.

4. Discuss the importance of reusing materials, and have the class develop a list of common items that can be reused and the new use(s) for each.

Cross-Curricular Integration

Home, safety, and career:

- If you have an industrial arts/woodworking shop in your school, have students construct inventions and artwork from scrap pieces of wood. This project could be team-taught with the technical education lab instructor. Invite parents or other resource personnel to work with the students.

Music:

- Have students research the history of tambourines, drums, maracas, and other musical instruments that ancient people may have fashioned out of used materials such as bones, pelts, seeds, and wood. Have students make their own instruments out of already-used materials.

Social studies:

- Have students research the lives of inventors and their inventions. Did any of them build their prototype inventions from discarded materials?

● Student Instructions for "Treasures from Trash"

Discarded materials lose their usefulness as resources when landfilled, and they can be dangerous to wildlife when they are not disposed of properly. Reusing materials in creative ways extends the "lives" of these materials. In this activity, you will design and construct your own invention from common household trash.

Safety and Disposal

Use caution when working with materials that have sharp edges. If you need to poke holes in a container, support the container from the inside with a chalkboard eraser as you poke holes through the sides. Or have your teacher or another adult punch the holes for you. Evaluate your designs for safety concerns and discuss safety with your teacher.

No special disposal procedures are required.

Procedure

1. Working with a group or by yourself, brainstorm several ways to use discarded items in inventions or artwork. For help in generating ideas, see the Student Background for information about machines, musical instruments, games, and artwork made from trash.

2. Choose from your list an item you would like to construct. It should be an item that is of interest to you and that is possible for you to construct with the materials provided. Before you begin working on your new creation, think about factors that will affect the design, like the class did with the bird feeder. Use a separate sheet of paper to answer the following questions and any others you feel will help you design your creation:
   • How will your creation be used? Where? When? By whom?
   • How will the pieces be fastened together?
   • What already-used materials might be helpful in making the creation?

3. Sketch your design on a separate piece of paper. Be sure to describe in detail how you will connect materials and how the pieces will fit together.

4. Discuss your design and the safety of your construction process with your teacher. Make appropriate changes and get final approval from your teacher before you proceed to step 5.

5. Using materials your teacher has provided and any items you have brought from home, build your creation. As you are building, you may find you need other materials. (If you must gather additional materials, try to select only already-used objects.) Keep track of all the materials you use.

6. Evaluate your finished creation by answering questions a–c in the Analysis Questions section of this handout.

Analysis Questions

a. Does your creation fulfill the functions you originally had in mind? How?

_____

_____

_____

b. How might you change the design to improve its function?

_____

_____

_____

_____

_____

c. If you were to mass-produce your creation, how might you change your design?

_____

_____

_____

_____

_____

_____

● Teacher Notes for "Shrinking Crafts"

In this activity students determine the percentage of change in dimensions (and mass) of a piece of polystyrene, a plastic that shrinks when heated. They will then design and create their own shrinking crafts from discarded polystyrene, extending the productive life of these "trash" materials and thereby reducing the waste stream.

Group Size .... 2 students
Time Required .... Getting Ready: 15 minutes
Procedure: 30–45 minutes

Safety and Disposal

You or another adult should be the only one to use the oven. Caution students about the dangers of a hot oven and hot items just removed from the oven. The polystyrene pieces will take a minute or so to cool completely after removal from the oven before they can be handled. Hot polystyrene can stick to fingers and cause painful burns. Caution students that rough edges on the plastic can be very sharp, especially after the plastic has been shrunk, and that they should make edges as smooth as possible before heating.

No special disposal procedures are required.

Materials

For the Procedure, Part A
Per class
- aluminum foil or cookie sheets

- 1 or more toaster ovens or access to a conventional oven
  Microwave ovens will not work.

- 2 hot pads or oven mitts
- kitchen spatula
- (optional) sandpaper

Per class or per group
- balance capable of measuring masses as small as 1g

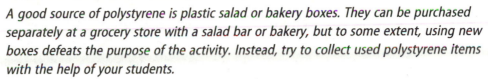

Per pair of students
- 5-cm x 7.5-cm rectangle made of paperboard or light cardboard
- piece of polystyrene (#6 recycle code) at least 10 cm x 10 cm

A good source of polystyrene is plastic salad or bakery boxes. They can be purchased separately at a grocery store with a salad bar or bakery, but to some extent, using new boxes defeats the purpose of the activity. Instead, try to collect used polystyrene items with the help of your students.

- scissors
- permanent marker
- ruler

For the Procedure, Part B

Per class
- aluminum foil or cookie sheets
- 1 or more toaster ovens or access to a conventional oven
  Microwave ovens will not work.

- 2 hot pads or oven mitts
- kitchen spatula
- (optional) sandpaper

Per class or per group
- hole punch
- (optional) needle-nose pliers (if making wire hook earrings or key chains)

Per pair of students
- 2–4 pieces of polystyrene (# 6 recycle code)
- scissors
- permanent markers of several different colors
- (optional) 1 or more of the following craft items from a local craft store:
  - earring posts or hooks
    Post earrings require glue, but wire hook earrings do not.

  - cords (for necklaces)
  - key chains
  - beads

Getting Ready

Cut out the 5-cm x 7.5-cm paperboard rectangles and mark an "L" for length on the 7.5-cm side and a "W" for width on the 5-cm side. If using a salad or bakery box as the source of plastic, cut out the flat surfaces.

Opening Strategy

Preheating the oven to 170°C (350°F) will ensure a quick shrinking time.

1. Ask students if they can think of examples of common trash items that could be reused instead of discarded. Discuss the term "property" and ask students how they think properties might determine whether an item could be reused and how it could be reused. Tell the students that they will investigate a property of a type of plastic called polystyrene and try to come up with another use for discarded polystyrene.

2. Hand out the Student Instructions for this activity and have the students do the activity. For older students, you may not want to include the Data Recording section but instead have them develop their own data table and formulas for the calculations.

If the curled edges of the hot plastic touch, they may stick. Should this happen, they can be pulled apart with extreme caution, using spatulas or oven mitts, while still warm and then reheated as needed to flatten. Either you should do this or the students should start over.

Cross-Curricular Integration

Art:

- Use shrinking plastics to make holiday ornaments, jewelry, key chains, or window ornaments.

Business, marketing, and economics:

- Have students make polystyrene products and sell them to other students in the school to raise money for a classroom purchase.

Mathematics:

- Have students predict the percentage by which the area of the shrunken plastic has changed. Have them calculate the area of the original piece of plastic, the area of the shrunken piece, the percent remaining, and the percent change in the area to test their predictions. (They can use the formulas from the Data Recording section of the Student Instructions.) Then, have students compare these numbers with the changes in length and width they found in the activity. Do they see any relationships between these measurements?

- For older students, have them cut out different geometric shapes such as circles, triangles, and hexagons from polystyrene. Students should measure the lengths of each side or the circumference of each shape before and after shrinking and find the percent remaining and the percent change. Have students calculate the areas of the original and shrunken pieces, the percent remaining, and the percent change. How do these calculations for other shapes compare to the calculations for a rectangle? Why might they be different?

Social studies:
- Investigate the history of plastics and how they have changed packaging and other aspects of human society. Students can also research how the use of plastics differs around the world.

Explanation

The shrinkable plastic used in this activity is polystyrene, a common polymer. Polystyrene (recycle code 6) is a rigid plastic that can be clear, as in salad containers, or foamed to improve its insulating properties in food service containers. Recycled polystyrene is used in many products, including office accessories, cafeteria trays, toys, video cassettes and cases, and insulation board. Depending upon how they are manufactured, polystyrene and certain other polymers can have the ability to shrink when heated.

The shrinking ability of polystyrene is somewhat unusual. Most common solids either expand before they melt into liquids (for example, metals) or decompose (for example, wood into charcoal). Polystyrene and other shrinkable plastics exhibit their shrinking nature due to the way they are manufactured. As they are produced, these plastics are heated, stretched out into a film, then quickly cooled. The sudden cooling freezes the molecules of the polymer in their stretched-out configuration. When the plastics are heated once again, the molecules within them are released from their frozen configurations; they return to their original dimensions, resulting in the observed shrinkage. Depending on how the polystyrene was stretched during manufacturing, it may not shrink uniformly when heated. For example, a circle may turn out to be an oval after shrinking. Students can expect a 25–75% reduction in length and width after shrinking, but little to no difference in mass.

Resource

Sarquis, J.L.; Sarquis, M.; Williams, J.P. "Shape Shifters"; *Teaching Chemistry with TOYS*; McGraw-Hill: New York, 1995; pp 151–160.

● Student Instructions for "Shrinking Crafts"

In this activity you will investigate the effect of heat on waste samples of plastic and apply what you learn to design and create your own shrinking crafts.

Procedure

Part A: Shrinking Plastic

1. Use the paperboard patterns provided to trace and cut out a rectangle of polystyrene.

2. Use a permanent marker to mark an "L" for length and a "W" for width on the plastic matching those on the cardboard pattern and write the names or initials of you and your partner on the plastic in large letters.

3. Measure and record the length, width, and mass of the plastic piece and record them in the "original" column of the Data for Shrinking Plastic table in the Data Recording section of this handout.

4. Place the plastic piece on aluminum foil or a cookie sheet and have your teacher heat it for about 15–30 seconds in a 170°C (350°F) oven. The shrinking is complete when no additional changes are observed for several seconds. When the piece is removed from the oven, let the plastic cool thoroughly (about a minute) before handling it.

   ➤ *Because oven temperatures vary, watch shrinking time carefully. Pieces of polystyrene will typically first curl and then lie flat as they shrink. (If the curled edges touch, they may stick. Discuss what to do about this with your teacher.) The plastic might need to be flattened while still warm with a smooth, flat item such as a spatula or piece of cardboard.*

5. Measure and record the length, width, and mass of the shrunken piece of plastic.

6. Calculate the percent remaining and the percentage of change in size and mass of the plastic sample using the formulas in the Data for Shrinking Plastic table.

7. Answer questions a–e in the Analysis Questions section of this handout.

Part B: The Challenge

1. Your challenge is to design a product that is made of discarded polystyrene. By reusing this material, you can keep it out of the solid waste stream. Remember that in order to be effective, your product must have two characteristics: 1) it should use up as much as possible of the piece of polystyrene and leave the smallest amount of waste possible; and 2) it should be marketable—that is, someone should want to use it or even buy it. Consider the properties (such as texture and appearance) of the plastic you shrank in Part A.

2. Discuss your design with your teacher before trying it out. If holes are needed in the final product, punch them before giving the plastic to your teacher to shrink. Because holes shrink with the rest of the product, punch holes about twice as large as they need to be to fit key chains or cords after shrinking. You may want to use sandpaper to smooth points or sharp edges on the shrunken pieces.

Data Recording

Table: Data for Shrinking Plastic

	original	shrunken	% remaining $\left(\dfrac{\text{shrunken size}}{\text{original size}} \times 100 \right)$	% change (100% − % remaining)
length				
width				
mass				

Analysis Questions

a. Compare the thickness of the shrunken piece with the thickness of a scrap of the original piece. What has happened to the thickness?

_____

_____

_____

b. Compare the percent remaining and the percentage of change of the length to those of the width. Are the percentages equal? If not, how are they different?

_____

_____

c. Did the mass of the plastic piece change after shrinking? What does this fact tell you about the shrinking process?

_____

_____

_____

_____

_____

d. How do you think the particular piece of plastic you were working with was created? Be sure to consider what happened to the thickness as well as the length, width, and mass.

_____

_____

_____

_____

e. Compare the appearance and properties of the shrunken plastic with a scrap of the original piece. How are they different? How are they similar? Would the shrunken plastic be suitable for use in the original product (such as a deli container or yogurt-cup lid)? Why or why not?

_____

_____

_____

_____

_____

_____

The Garbage Gazette

March 5   Local Edition   Vol. 1, Issue 8

Earthships Take Off

What is your home made of? Wood? Brick? Vinyl? Tires? Aluminum cans?

Tires and aluminum cans? For a house? The idea may seem strange to you, but to Michael Reynolds it's commonplace. He has been building homes from already-used materials for more than 25 years.

Reynolds' inspiration came in 1970 when he saw a television report about the beer-can litter problem in America. He thought there had to be a way to use cans and other discarded objects in building houses. Using his architectural training, he experimented with many designs until 1986, when he unveiled the "Earthship," a house built from cans, tires, and other already-used objects.

Tires are packed with dirt to form the "bricks" that are the basic structural unit of the Earthship. Whenever possible, these homes are built into a hillside, using nature to create some walls as the cave-dwelling Anasazi Indians of the southwest did. Once the tire bricks are in place, aluminum cans and more dirt fill in gaps. Interior walls are made of cans or bottles and plaster or concrete.

Earthships also reuse water. Rain water is collected and filtered for cooking, drinking, and bathing. "Gray water" from showers and sinks goes down the drain and circulates through interior flower and vegetable gardens. After it has nourished the gardens, the water is used in special toilets, either solar or composting, which then turn human waste into fertilizer for outdoor gardens.

Living in a pile of tires: the foundation of an Earthship is made of reused tires. (Photo courtesy of Earthship Global Operations, www.earthship.org.)

Earthships are especially popular in the Southwest, where the solar panels used to provide energy for the house can receive a maximum amount of sunshine every year. Whole neighborhoods of Earthships are being constructed in New Mexico and California, and Earthships have been built in Australia, Japan, and even on top of a 4,000-meter-high (13,123-foot-high) mountain in Bolivia.

Of all the materials used—or, rather, reused—to make Earthships, tires are perhaps the most important. They insulate the homes, keeping them cool in the summer and warm in winter. Tires also provide a strong foundation for the homes. Perhaps most important, by reusing tires in Earthships, people reduce the number of tires that must be disposed of.

About 250 million tires are discarded every year. In 1995, 69% of these were recovered for use as fuel, road fill, landfill cover, rubber products, agricultural applica-

tions, and other uses, including Earthships and artificial marine reefs. The other 31% are stockpiled or illegally dumped. If buried in a landfill, tires tend to rise to the surface, breaking open holes and disturbing contents, which may lead to landslides of trash. Standing tires provide an excellent breeding ground for rodents or mosquitoes that may carry diseases. Tire piles, if ignited either naturally or on purpose, are difficult to extinguish and release many chemicals into the environment, polluting the air, soil, and water.

Think About It

1. What kinds of materials are Earthships made out of?

2. Why is it important to dispose of or reuse tires properly?

3. What other materials can you think of that can be reused to build houses and buildings?

The Garbage Gazette

March 15   Local Edition   Vol. 1, Issue 9

Don't Dismiss Disposables

Suppose you walked into a McDonald's® restaurant and you were given the choice of a paper or glass container for your drink. The paper cup would be thrown away after your meal, while the glass cup would be washed for another customer to use. Which should you choose? Would the environment fare better if fast-food restaurants served food in reusable containers? The answer is not as simple as you might think.

Many people assume that disposable food containers have a more negative impact on the environment than reusable ones. This is a common idea, probably because disposable food containers create an impact we can easily see—cans full of trash. But to evaluate the whole environmental impact of reusable and disposable food containers, we must also look at their impact on water, air, and human health.

For example, consider the water waste created from the hot, soapy water needed to sufficiently clean a reusable cup after each use. In a 1992 Dutch study, researchers found that washing a porcelain or glass item even one time has a greater impact on water pollution than the manufacture of a one-use item. However, more air pollution is created during the manufacture of paper and plastic cups than those made of porcelain and glass.

Human health is another factor to consider. The disposable package used to wrap your hamburger or hold your french fries helps keep food safe

This glass won't get tossed in the trash, but it may be difficult to get it "to go."

from bacteria. The levels of bacteria found on reusable food-service ware like you would find in sit-down restaurants is higher than those found on disposable containers from fast-food restaurants. The acceptable bacterial level set by the U.S. National Academy of Sciences is fewer than 100 bacterial colonies per utensil. A 1993 study of a variety of food-service establishments found that reusable items contained a count above this level 17% of the time, while fast-food restaurants exceeded the limit only 3% of the time. A hygiene study in the state of Virginia found an average bacterial count of 410 colonies per reusable utensil compared with

an average of two colonies per disposable utensil.

When we consider all of the factors, choosing whether to use a disposable or reusable item is not easy. A 1996 review of more than 30 different studies on food-service packaging concluded that neither disposables nor reusables are consistently better for the environment. However, the review did draw these conclusions: 1. Disposable packaging is usually more hygienic than reusable. 2. Because it is not washed, disposable packaging is almost always better in terms of water use and wastewater production. 3. Reusable packaging may be better in terms of solid waste, air pollution, and energy use, but only if the item is reused at least several hundred times.

When choosing between reusable or disposable food packaging, health and environmental concerns are important, but we must consider other issues as well. Does a snack bar have the facilities to store and wash reusable dishware? Would disposable dishware be appropriate in a fancy restaurant? Paper cup or glass? You'll have to decide.

Think About It

1. *What factors need to be considered when deciding between reusable and disposable products?*

2. *What are some advantages of reusable products? disposable products?*

Lesson 6:
Resource Recovery

Resource recovery, in which useful materials or energy are removed from the waste stream prior to permanent disposal, is achieved by three major methods: composting, recycling, and incineration. Each topic is explored with one or two activities as well as Student Information sheets.

The investigation of composting consists of two activities. In "**Not Eggsactly Decomposing**," the decomposition rates of several different types of solid waste are tested by putting them into the wells of an egg carton with soil and water and making long-term observations. In the second activity, "**Compost Columns**," students create their own compost columns out of plastic 2-L bottles and watch decomposition in action.

Students explore some of the problems associated with recycling center separation in "**Now Separate It!**" by using the physical properties of different types of solid waste to separate them. In the second activity, "**Trash in the Newspaper**," students make their own paper from used paper and try to remove different types of paper contaminants.

The topic of incineration is covered in the activity "**How Good Is Your Fuel?**" in which students calculate the potential energy value of different items found in the waste stream.

What can recycled materials be used to make? *The Garbage Gazette* "**Fluff Up a Milk Jug for a Good Night's Sleep?**" looks at what products are made from recycled materials. The use of worms in composting, a growing trend, is explored in *The Garbage Gazette* "**Nature's Garbage Disposal**."

● Teacher Notes for "Not Eggsactly Decomposing"

In this activity your students will predict and then observe the decomposition rates of different materials.

Group Size.. 3–4 students
Time Required .... Procedure: 30 minutes
                                                                  Observation: 15 minutes per week for at least 10 weeks

Safety and Disposal

Caution students about sharp edges on some of the cut items. If using wooden matches, light and extinguish them beforehand to prevent students from lighting them. Instruct students to wash their hands thoroughly after handling the items in their egg cartons.

At the end of the observation period, have students separate out those items that did not decompose during the activity to be rinsed and reused, recycled, or thrown away. Put the soil and decomposed items on a garden or compost pile. Recycle the egg cartons if polystyrene recycling is available in your area.

Materials

For the Procedure
Per class
- (optional) balance
- (optional) plastic bag

Per group
- fine-point permanent black marker
- polystyrene egg carton with 12 wells

 Have your students help you collect polystyrene egg cartons a few days before doing the activity.

- small piece of each of the following:
  - newspaper
  - orange peel
  - aluminum foil
  - grass clippings or shredded leaves
  - cotton ball (natural cotton, not synthetic fiber)
  - wool yarn

- ◦ toothpick or used wood match
- ◦ cardboard
- ◦ glass marble
- ◦ iron nail
- ◦ metal paper clip
- ◦ plastic bottle cap
- ruler
- soil
- water
- wooden Popsicle™ stick, craft stick, or plastic spoon

Getting Ready

If necessary, cut or break the materials to be tested into pieces small enough to fit into the wells of the egg carton—about 1 inch x 1 inch. It may be helpful to place the test items in plastic bags to facilitate distribution.

Opening Strategy

1. Ask students whether they think a material left exposed in the outdoors for 10,000 years would retain its original form. *No.* What kinds of things would happen to it? Responses will depend on the materials chosen for discussion, but if necessary, bring up the idea that some metals will oxidize and that even rock can be worn away by the action of wind and water.

2. Define the term "decompose" (to break down more complex substances into simpler substances). Explain that decomposition is a chemical process that can be caused by temperature changes, light, electricity, or microorganisms. Other processes, such as weathering or rusting, can make an object become smaller in size but do not break down complex substances into simpler substances. In fact, rusting, which can cause material to flake off an iron object, actually involves creating a more complex chemical substance; rusting is an example of oxidation in which oxygen from the air combines with iron to form an iron oxide that we call "rust." Some other metals also form oxides that flake off.

3. As a class, come up with a list of common trash items. Ask your students which of these things would decompose and which would not if left exposed to the forces of nature.

4. Explain that students will conduct an experiment to determine what conditions promote decomposition. Discuss the kinds of conditions that common waste items are exposed to, such as soil, warmth or cold, rain, light or darkness, or sea

mist. (Other conditions, such as animals and trampling, are also valid, but they are probably not practical to test in the classroom.) As a class, make a list of several experimental conditions that can be tested. The list could include the following conditions:

- items in egg-carton wells with no soil (lid closed or open)
- items in wells with soil
- carton lid open in a sunny window or under a lamp
- carton in freezer

The items in the carton with no soil are the control; the class should decide whether the lid should be closed or open. All the rest of the cartons should have soil; however, if students bring up the idea of litter in waterways, they can fill one carton with fresh water and another with salt water, replenishing evaporated water as necessary.

5. Assign each group to test one of the conditions on the class list. Make sure to assign one group to prepare a "control" carton (no soil or water).

6. Hand out the materials and the Student Instructions for "Not Eggsactly Decomposing" and have the students do the activity.

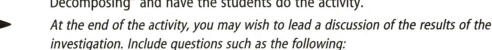

At the end of the activity, you may wish to lead a discussion of the results of the investigation. Include questions such as the following:
- *What types of materials decomposed readily?*
- *What types of materials did not change over the observation period?*
- *Do the students think these materials would change if observed for longer periods of time?*
- *Could the students make more accurate predictions now that they know more about what types of materials decompose and what types do not?*

Cross-Curricular Integration

Language arts:
- Have students keep written journals explaining their observations in more detail than the Data Recording section allows.

Life science:
- Discuss the concept of food chains and food webs and the importance of decomposers.

Mathematics:
- Have students graph the changes in size and mass of the 12 items over time.

Explanation

Decomposition is the process of separating pure substances into simpler compounds or elements. Decomposition can be caused by certain temperatures, light, electricity, or microorganisms. Decomposition by microorganisms is called biodegradation. In the presence of sufficient water and air, microorganisms can usually break down natural organic materials, which contain carbon and are derived from living organisms (such as orange peel or paper), more quickly than synthetic organic materials (like polystyrene or polyethylene plastics) or inorganic materials (such as glass or metal).

In this activity, the natural organic materials, such as the food and paper items, decompose the quickest. Although the items shrink in size and mass, the chemicals of which they are made do not disappear. They break down into solid, liquid, and gas particles that are too small to see. When the items are exposed to nature's forces, as is the case with litter, these particles become part of the soil, evaporate, or float away.

Items made of inorganic materials, such as glass and metal, will also change over time. The physical action of wind and water, for example, slowly removes tiny pieces of the item. This process is not the same as decomposition, because it merely involves breaking the item itself into smaller pieces, rather than breaking down the chemicals themselves. Chemical changes can also cause an object to break down without breaking down the chemicals themselves. Many metals react with oxygen to form oxides on their surface. Often the oxides flake off the surface, exposing more of the metal. Ultimately, the metal oxide becomes mixed with the soil. Other inorganic materials, such as carbonates, decompose when exposed to an acidic environment.

Some plastic products, such as trash bags, have been marketed as biodegradable, but in fact, many of these bags are made of small pieces of plastic held together with biodegradable starch. Once the starch decomposes, tiny pieces of plastic still remain intact in the environment. Recent research has begun to result in the manufacture of truly biodegradable plastics.

● Student Instructions for "Not Eggsactly Decomposing"

This activity allows you to observe the rate at which different trash materials decompose.

Safety and Disposal

Handle the trash items with care. Some of them may have rough edges. Always wash your hands thoroughly after handling the items in your egg carton.

At the end of the observation period, separate out those items that did not decompose during the activity to be rinsed and reused, recycled, or thrown away. If your teacher directs, sort out the soil and decomposed items for composting. Recycle the egg cartons if polystyrene recycling is available in your area.

Procedure

1. Use a permanent marker to make a 12-square grid on the inside of the lid of the egg carton corresponding to the 12 wells in the bottom part of the egg carton. In each square of this grid, write the name of one of the 12 test items. Also write your names on your group's carton.

2. Carefully observe the 12 items, recording their color, length and width, and mass (if a balance is available) in the appropriate columns of the Observations of Test Materials table in the Data Recording section of this handout.

3. Predict the rate of decomposition of each item on a scale of 1–12, with 1 being fastest and 12 slowest. Use an N to indicate a prediction that the item will not decompose. Record your predictions in the table.

4. According to your group's assigned experimental condition, place one of the 12 items in each well, being careful to match the labeled grid on the inside of the lid.

5. Once a week, "dig up" the wells with a craft stick or plastic spoon. Record the color, mass, and other noticeable characteristics of each piece. Then, carefully replace the items. Each week, start a new chart and write the number of that week in the "Week" heading of the table.

6. At the end of the observation period, dig up all of the wells and observe which items decomposed. Answer questions a–d in the Analysis Questions section of this handout.

Data Recording

Table: Observations of Test Materials

Item	Prediction (1–12 or N)	Week _____		
		Mass	Color	Other Observations
newspaper				
orange peel				
aluminum foil				
grass clippings/ shredded leaves				
cotton ball				
wool yarn				
wood				
cardboard				
iron nail				
glass marble				
metal paper clip				
plastic bottle cap				

Analysis Questions

a. Which materials showed change the soonest? (In other words, which materials had the fastest rate of change?)

_____

_____

b. What kinds of changes appeared first in the materials you listed for question a?

_____

_____

c. Which materials showed no changes after 5 weeks?

_____

_____

_____

d. Consider the items you tested. Now that you have observed the rates of decomposition for different materials, what recommendations could you make for waste disposal strategies for those materials?

_____

_____

_____

_____

_____

_____

● Teacher Background for "Compost Columns"

Composting is the controlled biological process of turning organic waste into a soil conditioner. In nature, organic matter is decomposed by bacteria and exposure to the elements, releasing nutrients that can be used by other organisms. Composting differs from natural decomposition in that conditions are controlled to make the decomposition occur more rapidly and efficiently. Composting produces a nutrient-rich soil additive called compost, which is used to improve soil quality (and thus plant growth) by increasing nutrient availability, water-holding capacity, aeration, and biological activity. Compost can be mixed into soil or applied on top of soil as a mulch.

Generally, only organic material of biological origin can be composted. These materials include wood, paper, animal waste, and plant materials. Breakdown is accomplished by a variety of microorganisms. Composting can be either anaerobic (occurring in the absence of oxygen) or aerobic (occurring in the presence of oxygen). The microorganisms that carry out anaerobic decomposition do not require oxygen, and the process does not produce heat. Since anaerobic composting occurs in the absence of oxygen, it usually is carried out in a closed vessel. This has the added advantage of keeping the foul odor from some of the decomposition products from escaping to the surroundings. Methane, a gas produced during anaerobic decomposition, can be collected and used as fuel. The end product of anaerobic decomposition is sludge, which can then be composted aerobically. An example of anaerobic composting is the treatment of human and household waste in a septic tank.

Aerobic composting is carried out by microorganisms that require oxygen. This process produces heat, which is beneficial because it destroys weed seeds and pathogens present in the organic material. When the pile is stirred regularly and is well ventilated, providing oxygen for the microorganisms, the products of aerobic decomposition do not cause odor problems. However, it can be difficult to maintain consistent conditions throughout the pile, so most aerobic systems have some anaerobic pockets.

Americans disposed of 29.75 million tons of yard waste—leaves, tree trimmings, and grass—in 1995, accounting for about 14.3% of the municipal waste stream. (EPA) Composting can reduce the amount of yard waste that must be landfilled or incinerated.

It is estimated that grass makes up about 50% of all yard waste. (EPA) This large volume of waste is unnecessary, because grass clippings do not usually need to be picked up. If grass clippings are short enough, they will quickly decompose and supply the soil with nitrogen and carbon. Longer grass clippings can be raked up and used as mulch. Composting yard waste reduces the volume of the original material by 50–85%. (Denison) Added to soil as a conditioner, compost improves texture, air circulation, and drainage.

Home composting can be an effective way to avoid both the expense of having waste removed and the expense of purchasing soil additives for the lawn and garden. However, if it is conducted improperly, home composting can be more of a hazard than it is worth. While a large pile that is kept moist and turned frequently can generate enough heat to kill the pathogens that can breed when food products (particularly meat and dairy products) decay, improperly tended compost piles can actually breed more pathogens, attract disease-carrying organisms like flies and rodents, and emit offensive odors. Experienced composters with plenty of land may be able to compost food waste without creating a nuisance, but beginning composters or people with small lots may need to limit themselves to yard waste and vegetable waste or simply not compost at all.

Despite the percentage of compostable waste that is regularly sent to landfills, centralized composting facilities have yet to catch on in many American communities for two major reasons: public image problems and economics. In the public's mind, composting facilities are malodorous nuisances. While properly-run composting operations produce little odor, it is difficult to prevent all anaerobic activity, so mild odor is emitted. Sealed systems and indoor facilities can help control the release of odor, but they increase cost substantially. In terms of economics, high-volume users (such as nurseries) and long growing seasons are needed in order for centralized composting facilities to distribute their product year-round. Compost is low in value and bulky. Prices must be kept low to attract customers, and most experts agree that the shipping distance from the composting facility to its market should be no more than 35 miles. Thus, municipalities should not expect to make money on composting—its purpose is cost avoidance. In fact, centralized composting may not prove to be economically practical at all for many communities; home composting may be a better alternative.

Besides its value as a method of waste reduction and soil treatment, composting has recently been discovered to have important applications in pollution prevention and control. Compost applied to creek, lake, or river embankments or on roadsides and hillsides reduces silting and erosion. It can also reduce heavy metals and organic contaminants in stormwater runoff, preventing contamination of water. Compost degrades or completely eliminates such contaminants as hydrocarbons and pesticides. In addition, the use of mature compost can suppress plant diseases.

● Teacher Notes for "Compost Columns"

In this activity, students will investigate the factors that influence decomposition in compost columns. They will design their own experiments, build their own compost columns, and observe the decomposition process over a period of 1–3 months.

Group Size.. 4–5 students
Time Required .... Getting Ready: 20 minutes
Procedure: 30–35 minutes
Observations: 10 minutes per week over the next 1–3 months

Safety and Disposal

Warn students that the cut edges of the plastic soft-drink bottles can be sharp. Show students how to use chalkboard erasers to support the bottles from the inside, and supervise closely as they poke holes through the sides of the bottles. For younger students, holes should be punched by an adult. In this case, you may wish to do steps 3a and 3b of the Student Instructions in advance.

After the observation period, the compost columns can be disassembled. The contents can be dumped on a garden or compost pile and the bottle parts can be rinsed out and recycled.

Materials

For Getting Ready
Per class
- hot water
- retractable blade, knife, or utility knife
- masking tape and pen for labels

For the Procedure
Per group
- pointed metal scissors
- nail
- clear tape
- measuring cups
- large spoon

Per student

- 3 clear plastic 2-L soft-drink bottles with 1 lid
- 10-cm x 10-cm piece of nylon stocking
- rubber band
- 2 cups soil
- 2 cups of 1 or more of the following kinds of organic matter
  - yard waste
  - newspaper or notebook paper
  - vegetable scraps

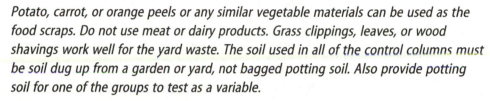

Potato, carrot, or orange peels or any similar vegetable materials can be used as the food scraps. Do not use meat or dairy products. Grass clippings, leaves, or wood shavings work well for the yard waste. The soil used in all of the control columns must be soil dug up from a garden or yard, not bagged potting soil. Also provide potting soil for one of the groups to test as a variable.

Per class

- 2–3 chalkboard erasers
- pitchers of water

For the Extensions

❶ Per class

- 2 empty compost columns
- balance
- 500 g shredded newspaper
- ½ cup soil
- water

❷ Per class

- several compost columns
- several different ingredients such as those used in the Procedure
- several identical sets of seedlings

Getting Ready

1. Prepare three plastic 2-L soft-drink bottles for each student: Pour about a cup of hot tap water into each bottle. Replace the cap, swish the hot water around in the bottle to soften the glue, and peel off the label.

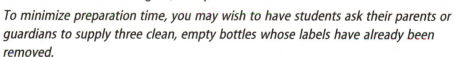

To minimize preparation time, you may wish to have students ask their parents or guardians to supply three clean, empty bottles whose labels have already been removed.

2. You or an adult helper should use the razor blade or a knife to make the initial cut in each bottle. Younger students may need lines drawn around the bottles for them to follow when they are cutting. Or you can do steps 3a and 3b of the Student Instructions in advance.

3. Number the bottles to help students follow the directions.

Opening Strategy

1. Explain to students that composting is controlled decomposition. Tell them that "decompose" means to break down into simpler compounds or elements, to disintegrate, or to rot. Ask students if any of them compost at home. What kinds of materials do they put in the compost pile? Where is the compost pile (enclosed in a bin or trash can, a pile on the ground, etc.)? Explain that several factors are important in composting and that students will investigate these factors by conducting controlled experiments with compost columns they will build themselves.

2. Introduce the concept of variables and the importance of testing only one variable at a time in an experiment. Explain to students that in order to determine the difference a particular variable makes in an experiment, scientists must have something to compare the result with. The object or process used for comparison is called the "control." The object or process in which one variable is altered is called "experimental." In this activity, groups will build control compost columns according to a basic plan and then build experimental columns that differ from the control by one variable. Each group will make at least two control columns and two experimental columns. Each column will be considered one trial. In an experiment, it is important to do more than one trial, in order to prevent unnoticed mistakes from providing misleading results.

3. Facilitate a discussion in which students brainstorm some possible variables in the composting process. The following are some variables that are important. If necessary, lead students to suggest these:
   - stirring
   - amount of soil
   - kind of soil
   - kind of waste
   - size of waste pieces
   - amount of water added

4. Divide the class into groups of four or five students. Assign each group one variable to test and have them design an experiment to test it. Have each group present its plan to you so you can review it for workability. Remind students

that at least two students in each group should build identical compost columns. Emphasize that within each group, the control columns should differ from the experimental columns in only one variable. Thus, students must ensure that all other conditions in their control and experimental columns are exactly the same. Discuss the importance of doing the following (unless testing one of these variables):

- Measure ingredients as exactly as possible.
- Use the same kinds of organic matter in all of your group's columns.
- Cut each kind of organic matter into pieces of uniform size.
- Put the ingredients in the compost columns in the same order.
- Keep the columns in the same location.
- Turn the compost using the same number of strokes.
- Pour exactly the same amount of water onto the columns the first time. (Afterward, use the water that has drained into the bottom of the column).

5. Distribute the Materials and the Student Instructions. Have students build their compost columns and observe them over a period of 1–3 months, recording their observations. (The observation chart in the Data Recording section of the Student Instructions contains space for 4 weeks. If the observation period lasts more than 1 month, distribute extra sheets as needed.) During this time, if a column starts to produce offensive odors, discuss the idea that a successful compost pile should produce minimal odor and that the experimental conditions in this column have prevented the column from achieving this goal. (Emphasize that this discovery is not the result of failure on the part of the column builder but is actually a desired result of the experiment, since it has demonstrated the importance of a variable in the composting process.) Then dispose of this column.

Discussion

1. When the observation period is finished and students have compared their final compost with others in the class, discuss the qualities of "good" compost. Observable qualities are as follows:
   - little or no smell
   - dark color
   - crumbly and slightly moist texture
   - homogeneous composition (Homogeneous composition means that all of the compost should look about the same throughout the mixture. Recognizable pieces of the original organic matter, such as paper or vegetable peelings, should not be visible.)

2. If possible, show students a sample of commercial compost and have them compare their compost with it. Discuss similarities and differences.

3. Discuss other qualities the students cannot observe directly, such as a beneficial balance of nutrients, soil pH appropriate for plant growth, and absence of weed seeds and pathogens. Discuss how students could determine these qualities.

4. Discuss differences between the compost columns and larger-scale compost piles. Differences include the size of the pile, the temperature that can be achieved, the amounts and kinds of waste added, and the monitoring of conditions in the pile. Discuss whether composting in 2-L bottles is an efficient strategy for producing large amounts of compost.

Extensions

1. Make a Newspaper Digester as follows: Use a balance to measure the mass of two equal quantities of newspaper (250 g each). Shred the paper and loosely pack one compost column with paper only. Mix ½ cup soil into the other batch of paper and loosely pack a second column. Pour equal amounts of water into each column and wait several minutes for it to seep through. If none comes out the bottom, add more in equal amounts until about ½ cup enters the reservoir. Pour the drippings back through the digester every few days for several weeks. Which digester decomposes faster? Why?

2. Make compost "tea," a liquid fertilizer, with compost columns. Filling several columns with different ingredients, such as some of those used in the Procedure, will result in drippings that are different in color and chemistry. For younger students, reinforce that this is tea for plants, not for humans, and it would be dangerous for them to taste it. Use the different liquids to water and fertilize identical sets of seedlings to see how different types of "tea" affect plant growth.

3. Discuss with students what would happen if plants and animals did not decompose when they died. What would we do with all of the bodies? How would nutrients get from those bodies into living plants and animals that need them to survive?

4. Have students build a compost heap with red worms. Information on building a compost heap with red worms can be found in *Worms Eat My Garbage*, by Mary Appelhof (Flower Press, #0942256107). You can also visit Mary Appelhof's World Wide Web site at http://www.wormwoman.com.

5. Challenge students to build their own compost bins or piles at home with parental permission as a final project or for extra credit.

Cross-Curricular Integration

Earth science:

- This activity easily complements a study of ecology or an investigation of environmental problems. Composting could lead into discussion of organic farming and farming that uses synthetic fertilizers.

Home, safety, and career:

- Have students investigate what happens to biodegradable materials in their community. Are residents allowed to put out yard waste with other trash, or must they take these wastes to a separate drop-off site? Does the community offer composting or chipping services for residents' compostable waste?

Language arts:

- Have students write stories about decomposition in nature or keep journals of their observations of their compost columns.

Life science:

- For older students, use this activity to lead into a discussion and investigation comparing "good" bacteria and nematodes (parasitic worms) to "bad" ones. Could some bacteria or nematodes be useful in one situation but harmful in another?

Explanation

Composting, or controlled decomposition of waste materials, is a biological process in which organic waste is broken down by microorganisms into a substance that is an effective soil additive. "Good" compost should be dark in color, moist, crumbly, relatively odorless, fairly uniform in texture, and rich in nutrients. Compost can be mixed into soil or applied on top of soil as a mulch.

Composting microorganisms, which include bacteria, fungi, and actinomycetes, come from soil that is added to the compost pile. If soil is not added to the pile, the necessary kinds of microorganisms will not be present to carry out all of the stages of aerobic decomposition. If no soil, or too little soil, is added, the final product will not have the texture of a good soil additive. While the waste material will still decay, it is likely to produce unpleasant odors, and it will not develop the texture of good compost.

Successful composting depends on nutrients, oxygen, pH, temperature, water, and physical properties of the composted material. The microorganisms in the pile require nutrients, which are supplied by the waste. Of the many nutrients involved, carbon and nitrogen are the major ones. Ideally, the ratio of available carbon to nitrogen should be between 25:1 and 30:1. Outside this range, composting proceeds more slowly. Excess carbon is converted to CO_2 and excess nitrogen is converted to ammonia until the balance is restored. Other nutrients, such as potassium, phosphorus, calcium, magnesium, and trace elements, are usually present in sufficient amounts.

Oxygen is required by aerobic microorganisms for respiration. Pores and air pockets exist throughout the pile. Some of these spaces contain water, but the rest are filled with air. Stirring and adequate ventilation help distribute oxygen to the microorganisms. Oxygen can also be replenished through diffusion and/or convection; as heat is produced through decomposition, warmer air rises and is replaced by cooler air drawn into the sides of the compost pile.

Desirable bacteria in an aerobic compost pile operate most efficiently when the pH ranges from 6.0 to 7.5. In the early stages, pH normally drops to about 4.5 or 5.0 as organic acids are formed. Then, as the acids are consumed, the pH rises to around 8, the normal pH of mature compost.

As mentioned previously, anaerobic decomposition produces heat. If left untouched, a large compost pile will increase in temperature to about 90°C. However, this temperature is too high for most bacteria. At temperatures above about 60°C, the number of bacteria decreases, and the composting process slows down. Temperature can be controlled by stirring the pile or forcing cool air through it. Another method is to make smaller piles, thus increasing the ratio of surface area to volume so that heat can escape more quickly from the center of the pile, where the temperature is the highest. The compost columns made by the students in this activity are so small that excessive heat is not likely to be a problem.

The microorganisms in a compost pile require water to survive. For greatest efficiency, the pile should be at least 45–50% (by mass) water. The maximum amount of water depends on the physical structure of the material. The coarser the material, the larger the pores and pockets, and the more water that can be retained without depriving the microorganisms of oxygen. In large compost piles, water tends to evaporate before it can penetrate far, so it is best to add water when the pile is being mixed. However, with the compost columns encased in 2-L bottles, rapid evaporation is not likely to be a problem. If the moisture level is high enough, liquid, called leachate, may drain from a compost pile. This leachate is likely to contain many nutrients, so it is beneficial to collect it and use it to re-water the pile as necessary.

The physical properties of the compost material determine whether the pile will have enough pores to hold a sufficient amount of air and water to support the microorganisms. Often, coarse materials such as wood chips or ground bark are added to a compost pile to increase pore volume. During composting, the pile tends to settle, reducing both the pile's total volume and its pore volume. Enough coarse material must be present to ensure adequate aeration throughout the decomposition process. While mixing restores pore volume temporarily, its primary purpose is to transfer material from the outside of the pile to the inside, where decomposition takes place.

● Student Instructions for "Compost Columns"

In this activity, you will work in groups to test different factors that are necessary to produce a successful compost pile. Each student in your group will build a compost column, and your group will test one variable.

Safety and Disposal

The cut edges of the plastic bottle can be sharp. Use chalkboard erasers to support the bottle from the inside while you poke the holes through the sides of the bottles.

After the observation period, compost columns can be disassembled. The contents can be dumped on a garden or compost pile, and the bottle parts can be rinsed out and recycled.

Procedure

1. As a group, design an experiment to test the variable you have been assigned to investigate. Keep in mind that each student in your group should build one compost column. Remember that in order for your experiment to be valid, you must keep all other conditions in your compost columns exactly the same.

2   Present your experimental design to your teacher for approval.

3. When your group's design has been approved, each student in the group should build a column according to steps a–g below. At least two students should build identical "control" columns, and at least two students should build identical experimental columns that differ from the control columns in only the variable being tested. Record the ingredients and conditions of your group's control columns and the experimental columns in the Data Recording section of this handout.

    a. Cut the top off Bottle 1 and the bottoms off Bottles 2 and 3 as shown in Figure 1.

    b. Holding a chalkboard eraser inside the bottle for support and protection, carefully poke ventilation holes in the top of Bottle 3 using pointed scissors or a nail. (See Figure 1.)

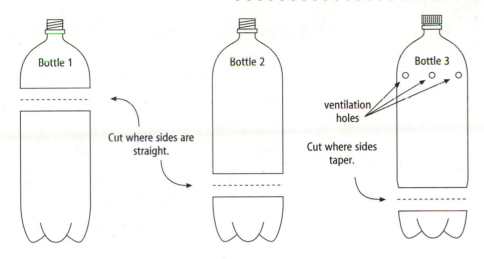

Figure 1: Cut the bottles as shown.

c. Place the piece of nylon stocking over the cap opening of Bottle 2 and secure it with a rubber band.

d. Assemble the compost column as shown in Figure 2.

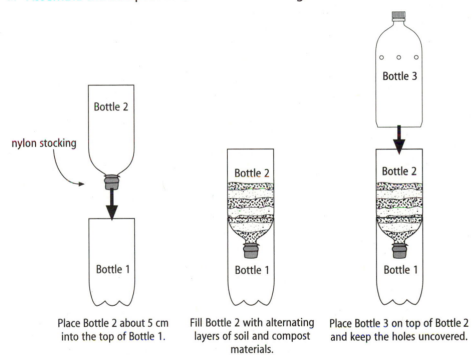

Place Bottle 2 about 5 cm into the top of Bottle 1.

Fill Bottle 2 with alternating layers of soil and compost materials.

Place Bottle 3 on top of Bottle 2 and keep the holes uncovered.

Figure 2: Assemble and fill the compost column.

e. If you are making a control column, starting with a layer of soil, place alternating layers of soil and compost materials (food scraps, yard waste, paper, etc.) into Bottle 2. If you are making an experimental column, fill Bottle 2 as described in your experimental plan. Record the materials you add to your column.

f. If you are making a control column, add just enough water to moisten the soil and allow a few drops to drain into the bottom of the column. If you are making an experimental column, add water as described in your experimental plan.

g. Place Bottle 3 in the top of Bottle 2, making sure the ventilation holes remain uncovered. (See Figure 2.) Be sure the cap is on Bottle 3.

4. Observe your group's compost columns over the next 1–3 months, turning the compost as scheduled in your experiment design. Record the appearance, odor, and any other important qualities of each column weekly on the Observations of the Compost Column table. Keep a separate table for each compost column, and record the compost column's number (for example, "Experimental 1" or "Control 2") in the heading of the second column of the table. (You will need a new table every month. Be sure to record week numbers in the first column of the table.)

5. At the end of the observation period, answer questions a–d in the Analysis Questions section of this handout.

Data Recording

Control Columns

Ingredients (in order) _____

_____

_____

Size of pieces _____

Location of columns _____

_____

Number of strokes to turn compost _____

Amount of water added to column (first time only) _____

Other conditions _____

_____

_____

_____

_____

_____

Experimental Columns

Ingredients (in order) _____

_____

_____

Size of pieces _____

Location of columns _____

_____

Number of strokes to turn compost _____

Amount of water added to column (first time only) _____

Other conditions _____

_____

_____

_____

_____

What effect do you think your variable will have on your compost column?

_____

_____

_____

Table: Observations of Compost Column

Week Number	Observations for Column _____

Analysis Questions

a. "Good" aerobic compost should have the following qualities: little or no smell, dark color, crumbly and slightly moist texture, and homogeneous composition. (Homogeneous composition means that all of the compost should look about the same throughout the mixture. Recognizable pieces of the original organic matter, such as paper or vegetable peelings, should not be visible.) Do the experimental columns in your group have any of these qualities? Explain.

_____

_____

_____

_____

_____

b. Do the control columns in your group have any of the qualities of good compost? Explain.

_____

_____

_____

_____

_____

c. Compare your group's compost columns with compost columns made by all of the other groups. Which column(s) produced the best compost in the class? How was this column made? Describe what qualities make its compost the best.

_____

_____

_____

_____

d. Consider the experimental columns made by all of the groups in the class. What do these columns tell you about the conditions necessary for making good compost?

_____

_____

_____

_____

_____

_____

_____

_____

_____

● Teacher Background for "Now Separate It!" and "Trash in the Newspaper"

Each person in the United States generates, on average, 4–5 pounds of municipal solid waste (MSW) per day. A large component of this waste can undergo resource recovery, and the materials can be salvaged as raw materials through recycling or composting or as energy through incineration. In 1995, after recovery for recycling and composting, the discard rate was 3.2 pounds per person (down from 3.3 pounds in 1994). (EPA)

According to the following Environmental Protection Agency estimates, annual recycling rates have steadily increased over the last three decades:
- 1960—5.61 million tons (6.4% of MSW)
- 1970—8.02 million tons (6.6%)
- 1980—14.52 million tons (9.6%)
- 1990—33.85 million tons (17.2%)
- 1995—56.19 million tons (27.0%)

Recycling can happen at two different times in the lifetime of the material. The first time happens after manufacturing, when the manufacturer collects scraps and leftovers and recycles them into new products. This is called pre-consumer recycling. The second time recycling can occur is after a consumer has bought an item, used it, and then sent it to be recycled. This type of recycling is called post-consumer recycling.

In 1994, Resource Conservation and Recovery Act (RCRA) Subtitle D regulations came into full force, imposing tighter regulations on every municipal solid waste landfill in the nation. In 1995, an estimated 27% of all solid waste was recovered, surpassing the agency's goal of 25% recovery by 1995. EPA projections indicate that a recovery rate of 30–35% is possible by the year 2000.

Much of our waste can be recycled or composted. Materials that are currently recycled in substantial quantities include office paper, magazines, plastic soft-drink bottles and milk jugs, glass containers, corrugated cardboard boxes, construction and demolition debris, wood, aluminum and other nonferrous metals, and iron and steel. Leaves, grass clippings, branches, and animal wastes are composted on a large scale. The following figure illustrates recycling rates of key household items in 1996. (EPA)

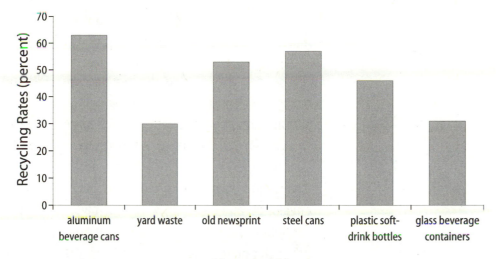

Recycling Rates of Key Household Items

In many cases, the cost of disposing of recyclable materials is higher than the cost of recycling. Aluminum is usually very profitable to recycle. The market for recycling newspaper was once a big money-maker for various organizations, but in many places it is now cheaper to landfill newspaper than to recycle it because the market is saturated. The cost of recycling versus the cost of landfilling varies from place to place. In some parts of California recycling can cost $40 per ton, whereas landfilling the same materials costs $60 per ton. In some parts of New Jersey, curbside recycling costs $50 per ton, whereas landfill disposal costs $100 per ton. Other estimates for disposal charges are $20–30 per ton for weekly curbside collection, $40–60 for landfilling costs, and costs of $90–110 per ton for incineration. (Selke) This savings does have exceptions: in some areas of the Midwest, curbside collection costs are nearly double the cost of landfill disposal. In such cases, more detailed analysis of costs and benefits is necessary to determine whether recycling (or a particular method of recycling) is beneficial.

Among the most easily recycled materials are paper, metals, glass, and plastics. The recycling potential of each resource is described in more detail below.

Paper

Paper and paperboard are recycled at a higher rate than any other material. In 1995, approximately 40% of the total paper and paperboard in the municipal solid waste stream was recycled. The largest single source of waste paper collected for recycling is corrugated cardboard boxes. In 1995, Americans recycled about 64.2% of the corrugated cardboard used, more than any other paper product. Stores and businesses use and recycle a steady stream of boxes. (EPA)

Most corrugated boxes now contain 20% recycled material; manufacturers predict this figure will soon rise to 40%. Recycling cardboard requires only about one quarter of the energy needed to manufacture cardboard from virgin materials and reduces the emission of air pollutants. (ESA)

Newspapers are recycled at a rate of 53%. (EPA) Top uses for recycled newsprint include more newsprint, paperboard, construction paper, insulation, egg cartons, and animal bedding. Old telephone books are another potentially large source of recyclable paper. Although they are made with the lowest possible quality paper, they can be reprocessed and made into ceiling tiles, textbook covers, and insulation. Mixed paper, such as colored paper, windowless envelopes, and cover stock, can be made into paperboard, the thin type of cardboard used in cereal boxes. Paper waste consisting of computer paper, office stationery, and other white paper is also made into roofing paper, tar paper, and asphalt shingles and can be sorted into grades of varying values.

Americans throw out about 85% of used white office paper (such as computer paper, photocopying paper, and adding-machine tape). Office paper is prized by recyclers because it is made with strong, short fibers that hold up well the second time around. Office paper has already been bleached and typically contains little ink that has to be removed. As a result, recyclers need to use only 25% as much bleach as is needed in the manufacture of virgin paper, which cuts down on emissions of dioxin, a by-product of paper milling. Making recycled paper uses an average of two-thirds less energy than pulping virgin trees. Water requirements can be as little as half that of making virgin paper, but if de-inking and bleaching are required, water savings are only about 15%. Recycling paper can reduce water pollution from papermaking by as much as 35% and air pollution by 74%. (Selke) One-third of the paper mills in the United States use waste paper exclusively.

Metals

Aluminum cans and other packaging are the largest sources of aluminum in municipal solid waste. Approximately 51.8% of all aluminum cans and packaging was recovered for recycling in 1995. (EPA) Using recycled aluminum cans saves 95% of the energy required to make aluminum cans from ore. Energy accounts for about 20% of the cost of producing aluminum from ore. Recycling aluminum represents an overall cost savings of about 40%. Recycling 1 metric ton of aluminum saves 4 metric tons of bauxite and about 700 kg of petroleum coke and pitch, and it also prevents the emission of almost 35 kg of toxic aluminum fluoride. Recycling aluminum reduces air pollution from the production of aluminum by an estimated 95% and water pollution by 97%. (Selke)

Americans use 100 million tin and steel cans every day. "Tin" cans are actually 99% steel, coated with a thin layer of tin to prevent rusting. (ESA) Recycling steel cans saves 60–70% of the energy used to produce them from raw materials and reduces air emissions by as much as 86%. Recycling also cuts down on water use and water pollution and conserves raw materials. (Selke)

Glass

According to EPA estimates, glass makes up about 6.2% of America's municipal solid waste. Over a ton of resources is saved for every ton of glass recycled: 1,330 pounds of sand, 433 pounds of soda ash, 433 pounds of limestone, and 151 pounds of feldspar. A ton of glass produced from raw materials creates 384 pounds of mining waste, while using 50% crushed recycled glass (cullet) cuts mining waste by about 75%. (ESA) Using 50% cullet can double the life of a furnace because cullet melts at a lower temperature than raw materials. (Selke) Recycling glass can reduce total air pollution by 14–20% and reduces water use by 50% and energy use by 25–32%. Most bottles and jars contain at least 25% recycled glass. (ESA)

Plastics

The overall recovery of plastics for recycling is only 5.3%, but recovery of some plastic containers is increasing. About 45.5% of all polyethylene terephthalate (PETE) soft-drink bottles and their base cups were recycled in the United States in 1995. (EPA) PETE bottles are actually a form of polyester. About one-third of all the carpeting made in the United States contains recycled PETE bottles. PETE can even be used to make clothing: 26 recycled bottles equals a polyester suit, and five recycled PETE bottles give enough fiberfill to make the insulation for a ski jacket. Recycling keeps about 360,000 tons of PETE out of landfills every year, but 1.33 million tons of PETE are still discarded (after recycling). (EPA)

High-density polyethylene (HDPE), a type of plastic commonly used in milk jugs, is recycled at a rate of 30.2%. (EPA) It can be recycled into items like flowerpots, trash cans, traffic barrier cones, and curbside recycling bins. Some detergent bottles are now made with 30% recycled HDPE. The only drawback to recycling HDPE is that controlling the color of recycled plastic is not possible, so it is sandwiched between layers of colorful new virgin plastic. According to experts, clean HDPE scrap can be worth $150–600 per metric ton; however, as much as $120 million worth of it is discarded every year.

Polyvinyl chloride (PVC) is used in pipe, fencing, and packaging. Currently, less than 0.05% of PVC in the waste stream is recycled. (EPA) The alternatives to recycling have important environmental drawbacks. In landfills, if certain PVC containers are exposed to water, solvents, or other trash, then the chemicals added to make the plastic more flexible (plasticizers) can leach into water and soil. PVC contains chlorinated compounds, so when it is incinerated with other trash, it releases hydrogen chloride, a corrosive gas.

Plastic bags and film wrappings account for about 20% of our plastic garbage. Although at present very little is recycled, this plastic has great potential for recycling. Plastic bags are among the few plastic products that may be recycled in a closed loop, meaning that recycled plastic bags can be made into more plastic bags. So-called "biodegradable" plastic bags do not eliminate the problem of plastic waste; sanitary landfills are designed to prevent decomposition, so landfilled "biodegradable" plastic bags do not biodegrade anyway. Also, many biodegradable plastic bags consist of plastic held together with cornstarch. The cornstarch degrades or dissolves, but the plastic itself does not. At best, the bags break into nearly invisible fragments. Thus, recycling is a better option than landfilling.

Recycling Programs

A number of different types of recycling programs can be implemented in a community:

- Drop-off centers at central locations accept recyclable materials from residents.

- Buyback and processing centers compensate individuals who bring in recyclable materials.

- Commercial recycling programs accept office paper, corrugated boxes, construction and demolition debris, bulk waste, and other materials.

- Residential curbside collection consists of gathering recyclable materials that residents have sorted from other trash items.

- Apartment building collections provide residents of an apartment complex with a centralized location to take their recyclables.

- Rural residents may be able to take their sorted recyclables to a landfill that has a container to collect those recyclables.

- Mixed-waste processing recovers recyclable materials from mixed solid waste. Material Recovery Facilities (MRFs), also known as Intermediate Processing Centers (IPCs), employ hand-sorting and various types of machinery, such as air classifiers, magnets, cyclones, trommels, crushers, grinders, and balers to produce clean, segregated loads of recyclable materials in large quantities.

● Student Background for "Now Separate It!" and "Trash in the Newspaper"

Today more people are recycling than ever before. We have all heard that recycling is good for the environment because it reduces landfill use and conserves the natural resources used to manufacture new bottles, paper, or cans. But is recycling as beneficial as we thought?

Many government agencies have set recycling targets. The U.S. Environmental Protection Agency set a recycling goal of 25% of all trash in the United States. This goal was surpassed in 1995 by a national recycling rate of 27%. Individual states have set their own recycling goals ranging from 20–70%. California and New York have both set state goals of over 50%.

But some people are questioning whether recycling quotas should be set that high and whether recycling is as beneficial as it is supposed to be. Twenty-five percent of our municipal solid waste (MSW) is made up of nonrecyclable materials like food waste, kitty litter, and dirt. To reach a goal of recycling 50% of all municipal solid waste, we'd have to recycle two thirds, or about 67%, of all recyclable objects.

It is possible, but difficult, to do so. Currently, about 65% of aluminum cans, 60% of paper and paperboard products, and 50% of steel from food cans and household appliances are recycled. Glass and plastic containers are also frequently recycled and turned into new products, like new bottles, insulation, plastic pipes, and building materials. So far, Americans are doing a good job with recycling, and the current rates are high. Do we need to increase recycling rates? Some argue yes, but others say no, because recycling isn't a perfect solution.

One reason many people became concerned about trash years ago was because of a fear we might run out of landfill space. In more crowded parts of the country such as the East Coast, this is a problem, but in other parts, there is enough room for landfills. Landfills are a good way to get rid of nonrecyclable solid waste, because they are carefully constructed to limit environmental hazards and they are usually located away from where many people live.

Some people are opposed to recycling because they believe it has just as much of a negative environmental impact as other disposal methods. One reason is that there are fewer recycling centers than landfills. Transporting recyclable materials to centers means using more fuel for trucks. Using more fuel could be harmful because fuel is a non-renewable resource, and its use causes air pollution. Another alternative to recycling is incineration, which involves burning solid waste to create energy. If solid waste is burned, then fewer natural resources, like coal and gas, are consumed.

Another problem is that recycling takes a lot of energy. The amount of energy used to clean glass bottles for reuse, to melt down plastic, and to remove ink from paper is sometimes higher than the amount of energy used to create those products in the first place. De-inking is especially troublesome because the sludge of old ink contains environmentally harmful heavy metals, and this sludge must be placed in landfills.

Another concern of recycling's opponents relates to paper. Paper companies are continually planting trees to produce enough wood pulp for paper. Because of this, the amount of timber in the United States is actually increasing. Some people worry that if we recycle enough paper, the paper companies won't need to plant trees and may sell the land to be developed for housing, shopping centers, or roads, which would permanently remove the trees.

There are many questions Americans need to ask themselves about recycling. No one would say it's completely wrong to recycle, but perhaps each community needs to think carefully about how much recycling it needs to do and what products are best to be recycled. Current programs work well, but it may be that they could work even better.

● Teacher Notes for "Now Separate It!"

Waste materials collected for recycling are often either mixed with other recyclables or with garbage. In either case, they must be separated prior to recycling. This activity challenges your students to develop a separation plan based on the physical properties of the materials to be separated.

Group Size .... 3–4 students
Time Required .... Getting Ready: 20 minutes
                                                              Procedure: 40–45 minutes

Safety and Disposal

Caution students about sharp edges on the cut plastic pieces. Each student should have a clean straw; straws should not be shared between students.

Items may be saved for reuse, recycled, or disposed of in the trash.

Materials

For the Procedure
Per group
- 2 identical sets of recyclable materials labeled "Test Materials" and "Final Process Materials" including the following materials:
  - 4–5 small pieces of clear polystyrene (PS), #6 plastic (from a clear salad container or yogurt lid)
  - 4–5 small pieces of high-density polyethylene (HDPE), #2 plastic (from milk jugs, shampoo bottles, etc.)
  - 4–5 small pieces of paper
  - 4–5 glass marbles or other glass objects without sharp edges
  - 4–5 aluminum foil pieces or aluminum lids from beverage containers (such as from glass bottles)
  - 2–3 steel or iron hex nuts or washers (must be attracted to a magnet)
  - 2–3 pieces of wood (such as Popsicle™ sticks or toothpicks)
- set of the following sorting equipment:
  - colander
  - magnet
  - 3–4 straws
  - scissors
  - bowl of water

- ○ 7 plastic cups to hold separated materials
- paper towels

For the Extensions
Per group
- all materials listed for the Procedure
- salt

Getting Ready

For each group, label one set of recyclables "Test Materials" and the other "Final Process Materials."

If you use only one set of materials for each group, be prepared to replace items such as the paper, which may get wet during the initial testing.

Opening Strategy

1. Ask the class to define physical properties. (A physical property is any property of an object that can be measured or observed without changing the identity of the substance.) Each substance has its own unique set of properties that make it different from any other material.

2. Ask the students to list different properties and write them on the board. Properties such as size, shape, color, density, magnetic attraction, strength, and solubility (ability to dissolve in a given liquid) can help to determine whether or not it is practical to recycle a substance.

3. Explain to students that they will be investigating some physical properties of different types of materials. They will use their findings to devise a way to separate one material from another so that individual materials could be recycled.

4. Give each group a set of Test Materials and a set of Final Process Materials, the sorting equipment, and the Student Instructions and have them begin their investigation.

When a group has completed their Proposed Separation Process table in step 3 of the Procedure, check it over briefly, comment as desired, and have the students conduct their separation plan using the Final Test Materials. If desired, lead a discussion of the various separation plans when students have finished their Procedure. Emphasize that there are many different, equally effective ways to separate the mixture. A community or recycling firm would tailor its plans to fit its needs and resources.

Extensions

1. Instead of providing the questions at the top of the Data Recording table, have the students come up with their own as they investigate the items. They may, for example, find that object size or static electricity could help in the separation process.

2. Have students change the density of the water by adding salt. How does this change the separation process? Could other separation methods be replaced if different densities of solutions were available?

3. Discuss the different separation methods used in recycling centers or have students do research projects on the different methods.

4. Tie this activity into other science lessons on topics such as physical properties, density, and magnetism.

Cross-Curricular Integration

Social studies:

- Discuss the history of recycling, including where, when, and why it has been done in the United States and how it has changed over the years. Compare recycling programs in this country with recycling programs in other countries.
- Have students investigate recycling in their home or school community. They could interview people who work at the recycling plant or community members about their views on recycling.

Explanation

Recycling can significantly reduce the amount of waste entering a landfill and also save resources. However, recycling facilities must be able to separate the different materials before they can recycle them. Each material can be identified by its own unique set of properties, including density and magnetism. Once students identify some of these properties, they will be able to design their own method of separation.

One method of separating metals is to use a magnet. Objects containing iron, such as a steel hex nut or washer, are readily attracted to a magnet.

When blown on through a straw, neither the wood nor glass moves far (marbles, if used, may roll but will remain in the colander), but the polystyrene, polyethylene, and paper can be blown out of the colander with low force, and the aluminum objects can be blown out of the colander with high force.

Density can also be used to separate materials. Other materials can be separated by their sink/float behavior in water. The ability of an object to sink or float in water is determined by its density (mass per unit volume) compared to the density of water. At room temperature (25°C), the density of water is 1.0 g/mL, the density of polystyrene (PS) is 1.05–1.06 g/mL, and the density of high-density polyethylene (HDPE) is 0.941–0.965 g/mL. Glass and PS will sink, while wood and HDPE will float. Paper will float initially but will sink if allowed to soak. The wet paper and PS can be separated by the fact that wet paper disintegrates if left in contact with water for a long time, and the solid pieces of PS can be filtered out.

Procedures used in actual recycling centers are similar to the ones developed by the students in this activity. For example, shredded plastics are dumped into a very low-density solution. Shreds that float are removed. The remaining shreds are placed into a slightly more dense solution and again the floating shreds are removed. This process is repeated over and over until all the plastics are separated.

● Student Instructions for "Now Separate It!"

Mixed waste arriving at a recycling plant must be sorted into categories before it can be recycled. In this activity, you will investigate the physical properties of common waste items and devise your own separation system based on these properties.

Safety and Disposal

Each student should have a clean straw; do not share straws. The plastic pieces may have sharp edges and should be handled with care.

Follow your teacher's instructions with regard to reuse, recycling, or disposal of waste items used in this investigation.

Procedure

1. Dump the Test Materials into the colander. Use the Material Properties table in the Data Recording section of this handout as a guide for testing the physical properties of each material in the mixture. Record all observations and discoveries. Dry the wet test materials (except for the paper), and return them to their bag. Discard the wet paper.

2. Design a separation process that would allow a recycling company to remove and recycle desirable items from the mixture you studied. Use your observations of the physical properties as a basis for design decisions. Separation techniques may not include separating the items by hand. However, after items are separated, they may be picked up and placed in the plastic cups.

3. Write your separation plan in the Separation Plan table. Share your plan with your teacher.

4. Test your separation plan on the Final Process Materials. You may wish to modify your plan as you learn more about the separation process.

5. Answer questions a–e in the Analysis Questions section of this handout.

6. (optional) Share your group's plan with the rest of the class. Evaluate any suggestions for improvement, keeping in mind that there are many "correct" ways to separate the mixture. Study other groups' plans and make your suggestions in turn.

Data Recording

Table: Material Properties

	Is the object attracted by the magnet?	Is the object blown by the straw? With how much force?	Can the object be cut easily with scissors?	Does the object float on water?	Would the object disintegrate if it is soaked in water?
plastic *yogurt lid* PS (#6)					
plastic *milk jug* HDPE (#2)					
paper					
glass *marble*					
aluminum *foil*					
metal objects *nuts, bolts*					
wood *toothpick*					

Table: Separation Plan

Step 1	
Step 2	
Step 3	
Step 4	
Step 5	

Analysis Questions

a. What are the steps in your separation process?

_____

_____

_____

_____

b. What methods worked well? What methods didn't work well?

_____

_____

_____

c. What difference does the order of steps in your process make?

_____

_____

_____

d. Do you think it's practical to recycle everything that is recyclable? Explain your answer.

_____

_____

e. Do you think energy costs connected to recycling are an important drawback, or do the benefits outweigh those costs?

_____

_____

● Teacher Notes for "Trash in the Newspaper"

In this activity, the students will make recycled paper and also investigate how to deal with contamination in the recycling process.

Group Size .... 4–5 students
Time Required .... Getting Ready: 20–30 minutes
                                                Procedure, Part A: 30 minutes + 5 minutes
                                                the next day
                                                Procedure, Part B: 30–40 minutes over
                                                2 days

Safety and Disposal

Use caution when operating the blender. The blender should be used only by an adult.

Caution students not to pour leftover slurry down the sink, as it will clog the drain. Flush the slurry down a toilet.

Materials

For Getting Ready
Per class
- scissors
- blender
- (optional) string

Per group
- 6 reusable wiping cloths, such as Handi Wipes® or Easy Wipes®
- 2 identical 2-piece wooden or plastic embroidery hoops 4–5 inches in diameter
- plastic window screen to fit hoop
- 2 double sheets of newspaper (1 double sheet is approximately 26 inches x 23 inches)
- oblong plastic dish pan
- water

For the Procedure, Part A

Per class

- (optional) "clotheslines" and clothespins (See Getting Ready.)

Per group of 4–5 students

- the following materials prepared in Getting Ready
  ◦ reusable wiping cloths cut in half
  ◦ dish pan filled with slurry
  ◦ mold
  ◦ deckle
- extra newspaper (for making the "couch")
- mixing spoon or stirring stick
- rolling pin, or pressing board made from 1 of the following:
  ◦ piece of wood approximately 9 inches x 12 inches
  ◦ hardback book inside a 1-gallon zipper-type plastic bag
- small pieces of paper and paper clips for labels

For the Procedure, Part B

Per group of 4–5 students

- all materials listed for Part A
- ¼ cup of one or more of the following "contaminants":
  ◦ vegetable oil
  ◦ plant matter, such as flowers and leaves
  ◦ coffee grounds
  ◦ crayon shavings
  ◦ soil
  ◦ Styrofoam™ pieces
  ◦ iron filings
  ◦ cornstarch

For the Extensions

Per class or group

- all materials listed for the Procedure, Part A
- beaker
- toilet paper
- (optional) office paper
- (optional) construction paper

Getting Ready

1. Cut the reusable wiping cloths in half to make rectangles approximately 10 inches x 13 inches.

2. (optional) Use string to make one or more "clotheslines" for hanging the paper up to dry.

3. Make a mold and deckle for each group as follows:

   a. Cut a square of window screen for each embroidery hoop with a length and width about 2 inches greater than the diameter of the inner ring of the embroidery hoops.

   b. Separate the rings of one embroidery hoop.

   c. Set the screen over the smaller ring and gently slide the larger ring into place (see Figure 1), keeping the screen taut. If the screen wrinkles, pull it firmly to smooth it.

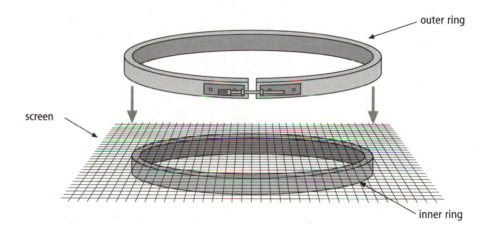

Figure 1: Gently slide the outer ring into place over the screen and inner ring.

   d. If the outer ring has a screw, tighten it. Trim the window screen to extend about half an inch outside the hoop. This is the mold.

   e. Keep the inner and outer rings of the second embroidery hoop together. (See Figure 2.) This is the deckle. It sits on top of the mold, with the hoops aligned with each other.

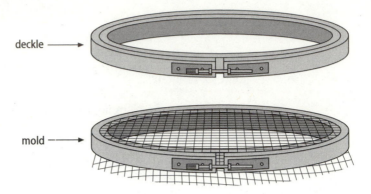

Figure 2: The mold and deckle go together.

4. Prepare batches of pulp ahead of time so that students do not have to use the blender. Make one dishpan of slurry for each group of students as follows:

   a. Tear the paper into pieces about 1 inch square.

   b. Fill the dish pan about half full of water.

   c. Put a handful of torn paper into the blender.

   d. Fill the blender about two-thirds full of water.

   e. Blend in 10-second bursts for a total of about 30 seconds.

   f. Pour the pulp into the dish pan of water.

   g. Repeat steps c–e three or four times more.

   Do not overblend the pulp; the fibers will become too short and weaken the paper. Short bursts of blending help prevent damage to the blender motor. Note that the final pulp mixture in the dish pan should be very watery—about ten times more water than paper. A common mistake when making paper is to use a pulp mixture that is much too thick.

Opening Strategy

1. Tell students that most paper is made from wood fibers derived from trees. The more paper we use, the more trees we must cut down. Also, paper is the largest single household waste component in landfills. However, the wood fibers in paper are reusable. Thus we can reduce solid waste substantially by recycling paper.

2. Tell the students that they are going to make recycled paper using a process similar to the commercial one but on a much smaller scale. Once they have

learned the technique, they will investigate the problems presented by contamination of the trash paper with other materials.

3. Introduce the words "mold" and "deckle," explaining that a mold is a frame with a screen that catches and drains the pulp, and a deckle is a matching frame with no screen that gives the paper its edge. Demonstrate how to make a mold and deckle using embroidery hoops and plastic window screen, as in Getting Ready, step 3.

4. Introduce the word "slurry." Explain that slurry is a mixture of paper pulp and water; you prepared slurry by adding water to small pieces of newspaper and mixing the solution in a blender.

5. Introduce the words "pulling" and "couching," explaining that pulling is gathering a sheet of pulp on the mold and couching is transferring the sheet of pulp to a cloth so it can dry. Demonstrate the steps in the Student Instructions for "Trash in the Newspaper."

6. Assign a contaminant for each group to work with in Part B of the Procedure, give the groups the materials and the Student Instructions, and have them begin the activity.

 If necessary, provide hints about how to separate contaminants in Part B. Some ideas are discussed in the Explanation section of these Teacher Notes. Students may become frustrated if they have been assigned contaminants that are difficult or impossible to remove. Use this opportunity to introduce the idea that recycling paper is much easier when the waste material is not contaminated and that a successful paper recycling program would be one that prevented contamination in the first place. Contamination of otherwise easily recyclable materials is one of the major disadvantages of Material Recovery Facilities (MRFs), which process unsorted waste.

Extensions

1. Recycle other types of paper. For example, stir toilet paper in a beaker of water to break it up into fibers. If a blender is available, use it to pulp other types of papers, such as office and construction paper.

2. Tie this activity into other science concepts such as dissolving, filtering, solutions vs. mixtures, and miscible vs. immiscible.

Explanation

Each person in the United States throws away nearly 4.3 pounds of unwanted materials daily. (EPA) Paper makes up about half of this waste and constitutes the largest single household waste component in landfills.

The materials in our environment that we use to make products are called raw materials, or natural resources. Natural resources include wood, oil, natural gas, coal, and minerals. When we make a product, we use up some of our natural resources. When we throw a product away, the natural resources used to make that item may be lost forever. The idea behind recycling is to use a product or material more than once, thereby extending the life of our natural resources.

Most paper, including notebook paper, computer paper, and newspaper, can be recycled. The single largest source of waste paper collected for recycling is corrugated cardboard. After a cardboard box has been used to deliver its contents, it is frequently recycled into a new cardboard box or other recycled paper products. Old newspapers are recycled into new newsprint and are combined with other scrap paper and recycled into boxboard, the familiar gray-colored board used to make boxes for cereal, soap, shoes, and tissue. These containers have a high content of recycled fibers, usually 90–100%. (ESA)

When trash is thrown away, it tends to be mixed together. This mixing can cause other materials to contaminate the paper in the waste stream. Many of these contaminants are difficult to remove, rendering the paper non-recyclable, but certain kinds of contaminants can be removed by various methods. When paper is recycled, it is mixed with water and pulped. At this point, different cleaning agents, such as bleach, are added to remove inks. Any contaminant on the paper will be mixed into the solution of paper and pulp. Contaminants such as Styrofoam and vegetable oil, which are lower in density than water, may float to the top and be skimmed off. Some water-soluble contaminants, such as food color, may be removed during the bleaching process. Solid contaminants, such as food particles, are more difficult to remove. They can be filtered out from the paper if they are large enough, but it would be difficult to come up with a screen fine enough to trap small particles of contaminants but large enough to allow the paper pulp to go through.

● Student Instructions for "Trash in the Newspaper"

In this activity, you will make recycled paper using methods similar to those used by manufacturers. You will also investigate how contamination can affect the paper recycling process.

Safety and Disposal

Do not pour leftover slurry down the sink, as it will clog the drain. Flush the slurry down a toilet.

Procedure

Part A: Papermaking

You will work in groups of 4–5.

1. Place a clean, damp, reusable wiping cloth on a small stack of newspaper. Roll up several additional sheets of newspaper into a cylinder and put the cylinder underneath the first stack of newspaper to prop up the stack at one end. This cloth-covered stack of paper, called a couch, should now form a gentle slope. Position the couch with the elevated edge near you. (See Figure 1.)

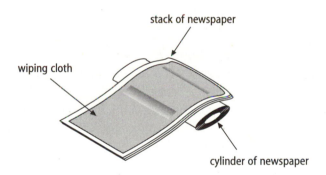

stack of newspaper

wiping cloth

cylinder of newspaper

Figure 1: Assemble the couch.

2. Vigorously stir up the pulp mixture in the dish pan with the mixing spoon or stirring stick to prevent the fibers from settling on the bottom of the dish pan.

3. Hold the mold and deckle together with both hands, keeping the deckle on top and the screen side of the mold facing up. Dip the mold and deckle into the pulp mixture at about a 45° angle, and then level them (while they are still in the pulp mixture). (See Figure 2a.)

4. Keeping the mold and deckle level, lift them straight up out of the pulp mixture and let the water drain away. (See Figure 2b.) While the pulp is still wet, gently shake the mold and deckle to help the fibers settle. Carefully lift off the deckle and set it aside.

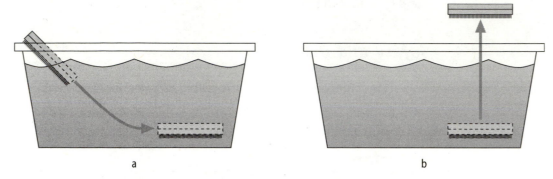

a   b

Figure 2: Pulling pulp involves (a) dipping the mold and deckle into the pulp mixture at a 45° angle and leveling them, and (b) lifting the mold and deckle straight up out of the mixture.

5. Place the mold on its edge at the top of the slope on the couch, with the pulp facing down the slope. (The pulp will not fall off the screen.) Carefully tilt the mold down the slope (away from you) until it lies screen down on the cloth. Gently press the screen against the cloth. Lift the mold slowly; the pulp will remain on the cloth. Label your creation by writing your name on a small slip of paper and attaching it to the cloth with a paper clip. Place another cloth over the pulp. (See Figure 3.)

If the sheet is too thin (transparent or falling apart), you may not have dipped your mold and deckle deeply enough in the dish pan. Try dipping it almost to the bottom before bringing it up. If this doesn't work, ask your teacher to make another blender full of pulp and add it to the dish pan. If the paper sticks to the mold, scrape up the pulp and put it back in the dish pan.

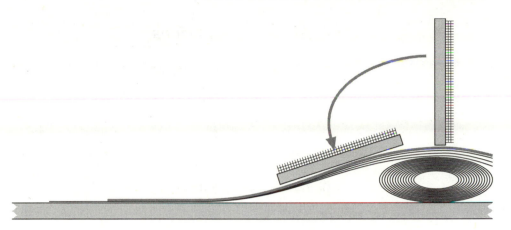

Figure 3: Couch the paper by tilting the mold until it rests on the cloth. Gently press the screen against the cloth and slowly lift the mold.

6. When everyone in your group has made a sheet of paper according to the instructions in steps 2–5, end the stack with a cloth and then a small stack of newspaper. Remove the rolled-up newspaper. Press the whole stack under the pressing board or roll with a rolling pin.

7. Carefully peel off each layer (cloth and the pulp sheet on top of it together) and clip it to the "clothesline" or lay it out to dry overnight. The paper will peel off the cloth when dry.

Part B: Contamination

1. Repeat Part A, adding about ¼ cup of contaminant in step 2 of Part A. Observe the paper made with the contaminant and answer questions a–c in the Analysis Questions section of this handout.

2. Discuss the contamination-removal method you proposed in question b with your teacher and then try it.

3. Compare the paper you made in Part A with the paper from steps 1 and 2 above and answer questions d–g in the Analysis Questions section.

Analysis Questions

a. What was your contaminant?

_____

_____

b. How is the paper made with the contaminant different from the paper you made in Part A?

_____

_____

_____

_____

c. How might you remove the contaminant before making the paper?

_____

_____

_____

d. Did any method(s) successfully remove the contaminant?

_____

_____

e. Would removing the contaminant before the water was added to the paper be easier? Explain.

_____

_____

_____

_____

f. Considering the results of all groups in the class, do you think it is always practical to recycle paper? What conditions might be necessary to make paper recycling practical?

_____

_____

_____

_____

_____

g. Consider the different types of recycling programs that can be used in a community, such as drop-off centers, curbside collection, and mixed-waste processing. In terms of what you learned in Part B, which strategy or strategies might be the best for collecting paper to recycle? Why?

_____

_____

_____

_____

_____

_____

_____

● Teacher Background for "How Good Is Your Fuel?"

In addition to recycling and composting, resources can also be recovered from the municipal waste stream through the process of incineration. In addition, incineration increases the useful life of available landfills and minimizes odor and sanitation problems. An efficient mass-burn incinerator can reduce the solid waste going into a landfill by as much as 80–90% in volume and 65–75% in mass. (Denison)

The most common type of incineration, mass burn, is designed to burn virtually all the waste brought to it, with no separation or processing of materials prior to burning. Most mass-burn facilities operating today include an energy recovery system that converts heat from the combustion process into steam or electricity that can be used by the surrounding community. The basic mass-burn process is outlined in the six steps below and shown in Figure 1.

1. Transportation—Municipal solid waste is collected and delivered to the mass-burn facility.

2. Storage—Waste is transferred to a storage pit or tipping floor.

3. Combustion—A conveyor or crane transfers the waste to the hopper, which feeds the waste into the furnace. Secondary combustion chambers aid complete combustion.

4. Energy recovery—The heat from combustion is transferred to water in pipes, which turns into steam. Steam is used directly for processes or to generate electricity.

5. Emission control—Dry and wet scrubbers and other air pollution control devices, such as electrostatic precipitators and fabric filters, remove acid gases and particulates from the exhaust.

6. Disposal of residue—The ash from burning, as well as residue from scrubbers and other pollution control devices, is disposed of in a landfill.

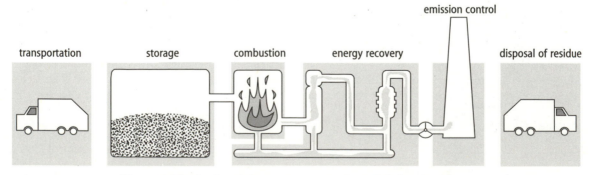

Figure 1: The incineration process consists of six basic steps.

Understanding Garbage and Our Environment

Waste that goes through the incineration process exits the system in one of four forms:

- Combustion gases—exit through the stack or may be removed by air pollution control devices.

- Particulate emissions—lightweight particles that exit the combustion chamber along with combustion gases and are small enough to get past pollution control devices.

- Fly ash—particles that are light enough to be borne upward with combustion gases but heavy enough to fall or large enough to be captured by pollution control devices before exiting the stack; approximately 25% of all incinerator ash; often contains high levels of heavy metals, acid gas constituents, and products of incomplete combustion (PICs), such as dioxin; is considered toxic.

- Bottom ash—uncombusted waste, such as glass and metal, generally considered non-toxic, approximately 75% of all incinerator ash

Most energy recovery facilities use sophisticated combustion control systems designed to optimize combustion, minimize ash for disposal, and help waste burn cleaner by reducing the formation of products of incomplete combustion (PICs).

The energy content of different kinds of solid waste varies. Paper accounts for more than 50% of the energy content and plastics for nearly 25%. (Selke) Plastics, which are derived from natural gas and petroleum, have a stored energy value higher than any other material commonly found in the waste stream. When plastics burn, they help other wastes combust more cleanly and completely. A pound of mixed municipal solid waste contains approximately 4,800 British thermal units (Btu) of heat, compared to about 13,500 Btu for a pound of anthracite coal. (EPA) In 1993, U.S. refuse-to-energy plants produced approximately 328 trillion Btu, and the EPA projects that amount will increase to 2.4 quadrillion Btu by the year 2000. (NREL)

A disadvantage of incineration is that not all waste is suitable for combustion. One quarter of the waste stream (by mass) is not suitable for incineration, including construction and demolition debris, and bulk waste, such as discarded stoves, refrigerators, and furniture. Preprocessing and removing unsuitable waste before combustion can dramatically reduce the amount of ash as well as the toxicity of ash and emissions.

The most obvious way humans are exposed to by-products of incineration is through the inhalation of gases or particles released into the atmosphere. Deposition on the ground, irrigation, runoff, crop ingestion, animal product ingestion, or direct exposure to deposited materials are also possible sources of exposure to airborne emissions. Less obvious is the potential for airborne and waterborne dispersion of ash during storage, transport, and handling for final disposal at the landfill. Yet despite concerns about ash toxicity, there is widespread interest in recycling bottom ash, especially for use as aggregate in concrete or cement.

Other disadvantages of incineration include large capital investment, relatively high operating costs, the expense of sophisticated pollution control equipment, and difficulty in obtaining sites. Studies show that while the majority of Americans support incineration as a national disposal option, they reject the concept of such a facility in their own community.

● Teacher Notes for "How Good Is Your Fuel?"

In this activity, your students will predict whether or not given items will burn. They will then try to burn each item and measure the amount of energy given off by those that do burn.

Group Size .... 3–4 for older students, or as a teacher demonstration for younger students

Time Required .... Getting Ready: 20 minutes
Procedure: 40–50 minutes

Safety and Disposal

Be sure your students use standard safety measures when working with matches or open flames. Consult local fire safety regulations for the room you will be using. Be sure the room has adequate ventilation and that a fire extinguisher is immediately accessible.

Use alcohol or metal cooking thermometers in this experiment. Avoid mercury thermometers because of the potential for breakage and the toxic nature of mercury.

Carefully examine the items to be burned for colorless plastic coatings, which may release harmful vapors when burned. Plastics should be burned only in a fume hood or outside.

Allow all of the burned materials to cool thoroughly, then dispose of them in the trash when the activity is finished. Materials that do not burn may be reused.

Materials

For the Procedure
Per class
- electronic or triple-beam balance
- fire extinguisher

Per group
- 100-mL graduated cylinder or metric measuring cup
- water
- soft-drink can

- 3-pronged clamp with 6-cm grip
- ring stand
- alcohol or metal cooking thermometer
- metal paper clip stand prepared in Getting Ready
- 2.5-cm x 2.5-cm piece of 1 or more of the following "trash" items (unless another size is indicated):
  - newspaper
  - dry crackers, such as saltines
  - aluminum foil
  - waxed paper
  - cotton ball
  - cardboard
  - magazine paper
  - 15-cm piece of wool yarn
  - thin pieces of balsa wood or a 2.5-cm length of a craft stick
  - glass marble
  - metal paper clip
  - plastic

 Burn plastic only outside or in a fume hood.

- commercial aluminum weighing tray or weighing tray made from heavy-duty aluminum foil (See Getting Ready.)
- matches

Per student
- goggles

Getting Ready

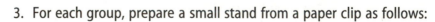

1. Thoroughly rinse out the soft-drink cans.

2. If using heavy-duty aluminum foil in place of a commercial aluminum weighing tray, shape the foil into a small tray (about 3 inches x 3 inches) with low edges.
   Use caution when shaping the foil; its edges will be sharp. Fold the sharp edges to the inside of the tray.

3. For each group, prepare a small stand from a paper clip as follows:

   a. Bend the outermost end of the paper clip out to the side at a 45° angle. This will be the base of the stand. (See Figure 1.)

   b. Bend the remaining inner hook upward at a 45° angle up from the plane of the base.

c. Bend the end segment of the small hook so that it is vertical.

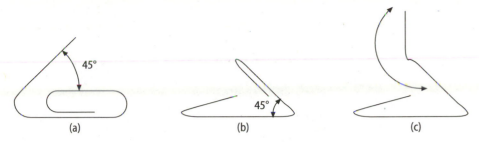

Figure 1: Make a paper clip stand.

Opening Strategy

➤ *For younger students, this activity works best as a teacher demonstration with students making predictions, recording the data, and doing the calculations.*

1. Ask students, "What is needed to make something burn?" *Fuel, oxygen, and heat (or a spark).* Show students the items to be tested in the activity. Have them predict whether or not the items will burn. Have them record their predictions in the Predictions and Data for Burning table in the Data Recording section of the Student Instructions. Is energy stored inside the items? Which items will give off the most energy as heat? Which will give off the least?

2. Give each group one or several of the "trash" items and the Student Instructions. Have them follow the Procedure.

➤ *It is not necessary to give each group all of the trash items; the number you give will depend on the time and resources available to the class. It is important, however, to make sure every group has at least one item that will burn.*

Extensions

1. Discuss with students the items that did not burn in this experiment. Do students think there is a way to make them burn? Would they melt? What effect would high temperatures have on them?

2. Ask students what effects they think liquids would have on the burning ability of the waste stream. How might different mixtures of items in the waste stream affect burn temperatures, energy produced, and the amount and type of ash?

3. Have students design an incineration plant. What kinds of factors would they have to take into consideration? What types of waste would the facility accept? Besides the ash, what other by-products of incineration would they have to deal with?

4. This activity can be used in discussions of melting points, chemical and physical properties, combustion, energy, and environmental science.

Cross-Curricular Integration

Social studies:

- Have students research laws about smokestack emissions or other environmental regulations.

Explanation

The main products of burning in this activity are carbon dioxide, carbon, water, and heat. Most of the heat energy produced when a test item is burned is absorbed by the water in the can mounted above it. In the process, the temperature of the water increases. One calorie is defined as the amount of heat necessary to raise the temperature of 1 g of water 1°C. By measuring the temperature increase in the water, students can calculate approximately how much heat energy the item produced when it burned (or underwent combustion).

The number of calories produced can be determined by multiplying the mass of water by the density of water by the change in temperature (in degrees Celsius) and then by the specific heat of water. For this activity, the mass of water is 100 g (100 mL of water has a mass of 100 g). The specific heat of water is 1 calorie/g-degree. To calculate the number of calories per gram of item burned, the mass of the item must be determined before burning. To do this, it is necessary to measure the mass of the items and tray before burning. In this activity, the black carbon-based residue left behind on the tray and on the bottom of the can is a result of the temperature not being high enough and oxygen at the point of contact being insufficient to completely convert all the carbon to carbon dioxide.

As with most experiments, several sources of experimental error are present in this activity. For example, not all of the heat from the burning is used to raise the temperature of the water. Some of the heat energy is lost to the surrounding air, while some is absorbed by the can. This loss of heat causes the calculated caloric value to be lower than the actual value.

Commercial incinerators are much more efficient than the setup used in this activity, and much of the energy value of municipal solid waste (MSW) can be recovered through waste-to-energy incineration. Modern energy recovery facilities burn MSW in special combustion chambers and use the resulting heat energy to generate steam. This steam can either be used directly or used to generate electricity through a steam turbine. Approximately 2.5–3 pounds of steam can be produced per pound of solid waste incinerated. Each ton of waste produces 400 to 500 kilowatts of energy per hour. As a potential fuel source, incineration could accommodate 2.6% of the nation's energy needs. (ODNR)

● Student Instructions for "How Good Is Your Fuel?"

In this activity, you will predict whether or not given items will burn, and then you will test these items to see how much energy is produced when each burns.

Safety and Disposal

Make sure that you follow your teacher's instructions when using matches or open flames and that you know where to find the fire extinguisher.

Carefully examine the items to be burned for colorless plastic coatings, which may release harmful vapors when burned. Plastics should be burned only in a fume hood or outside.

Allow all of the burned materials to cool thoroughly, then dispose of them in the trash when the activity is finished. Materials that do not burn may be reused.

Procedure

1. Use a graduated cylinder or measuring cup to measure 100 mL tap water. Pour the water into the soft-drink can through the open top.

2. Fasten the can into the 3-prong clamp on the ring stand.

3. Place the thermometer in the can, and read and record the initial temperature of the water.

4. Insert the straight vertical end of the paper-clip stand directly into your item if the item is soft. For firmer items, you may need to bend the vertical end of the paper clip into a small loop that you can set the item on. Place the stand and item to be burned on the aluminum tray.

5. Measure the mass of the tray containing the stand and test item and record this as the initial mass in the Predictions and Data for Burning table in the Data Recording section of this handout.

6. Place the tray containing the paper-clip stand and test item on the base of the ring stand underneath the can. Position the clamp so that the bottom of the can is 4–5 cm above the item to be burned. Arrange the setup so the test item is centered directly under the suspended can.

 Proper safety measures should be used when working with an open flame. Use a flame-resistant surface and remove unnecessary flammable materials from the area. Long-haired people should tie their hair back when working near an open flame.

7. Put on your goggles, then light the item with a match and allow it to burn.

8. When the item has finished burning, gently stir the water in the can with the thermometer and note the highest temperature the water reaches. Do not let the thermometer rest on the bottom of the can, as this will artificially inflate the temperature reading; the temperature of the water, not the heated can surface, is needed. Record this temperature as the final temperature.

9. Calculate the rise in temperature of the water in the can (final temperature minus initial temperature), and record this as the temperature change.

10. Using the following formulas, determine the heat produced (in calories) and the heat-to-mass ratio (in calories per gram) for each item burned. Record these values in the Energy Produced from Burning table in the Data Recording section.

$$heat\ produced = 100\ mL\ water \times density\ of\ water \times rise\ in\ temperature\ (in\ °C) \times specific\ heat\ of\ water,\ where$$

$$density\ of\ water = 1\ g / mL$$

$$specific\ heat\ of\ water = \frac{1\ calorie}{1\ g \times 1°C}$$

$$\frac{heat}{mass} = \frac{heat\ produced}{mass} = calories\ per\ gram$$

11. Repeat steps 3–10 for each item your group is testing.

12. Share your results with other groups and add their information to your chart.

13. Answer questions a–b in the Analysis Questions section of this handout and share your answers with the class.

Data Recording

Table: Predictions and Data for Burning

Item	Prediction: Will It Burn?	Mass (g)	Temperature (°C)		
			Initial	Final	Change
newspaper					
dry cracker					
aluminum foil					
waxed paper					
cotton ball					
wool yarn					
wood					
cardboard					
magazine paper					
glass marble					
metal paper clip					
plastic piece (record type) _____					

Table: Energy Produced from Burning

Item	Heat Produced (calories)	Heat/Mass Ratio (calories per gram)
newspaper		
dry cracker		
aluminum foil		
waxed paper		
cotton ball		
wool yarn		
wood		
cardboard		
magazine paper		
glass marble		
metal paper clip		
plastic piece		

Analysis Questions

a. How do the heat/mass ratios of the items your class tested compare to the heat/mass ratios of items (listed in the following chart) commonly burned in waste-to-energy facilities? Why do you think that some of the items your class tested are not commonly burned for energy?

_____

_____

_____

_____

_____

_____

Equivalent Fuel Values					
Component	Heat/Mass Ratio		Component	Heat/Mass Ratio	
	cal/gram	Btu/lb		cal/gram	Btu/lb
Coal (Anthracite)	7,500	13,500	Waxed milk cartons	6,292	11,325
Coal (Bituminous)	7,778	14,000	Polyethylene	10,382	18,687
Peat	2,000	3,600	Polystyrene	9,122	16,419
#2 Fuel oil	10,000	18,000	Mixed plastic	7,833	14,100
Mixed MSW	2,667	4,800	Tires	7,667	13,800
Mixed paper	3,778	6,800	Leaves (50% moisture)	1,964	3,535
Newsprint	4,417	7,950	Leaves (10% moisture)	4,436	7,984
Corrugated cardboard	3,913	7,043	Grass (65% moisture)	1,494	2,690
Magazines	2,917	5,250	Green wood	1,167	2,100
Mixed food waste	1,317	2,370	Cured lumber	4,056	7,300

b. Why do you think a waste-to-energy facility would choose to burn a material like coal instead of a material like polyethylene, even though coal has a lower heat/mass ratio?

_____

_____

_____

_____

_____

_____

The Garbage Gazette

April 14   Local Edition   Vol. 1, Issue 10

Fluff Up a Milk Jug for a Good Night's Sleep?

You drink out of plastic bottles, but would you sleep on one? Would you wear one? Check the tags on your clothing and pillows, because you could be doing both without knowing it.

Using discarded products (such as plastic bottles) to make new products is called recycling. Recycling is an important waste management option because it turns discarded products back into their original components, preventing them from being sent to landfills or incinerators. Thus, recycling recovers resources and saves time, energy, and money.

Products can be made of two different types of recycled material: post-consumer waste and pre-consumer waste. When you put used paper into a recycle bin, it becomes post-consumer waste because it was used by you, a consumer. When a manufacturer collects scrap pieces of aluminum left over after cans are manufactured and uses it to make new cans, the aluminum is pre-consumer waste because it was recycled before you, the consumer, had a chance to use it. Labels for recycled products list the amount of post- and pre-consumer waste used to make the products.

In the past, manufacturers simply made the same product from the recovered materials. Soft-drink bottles were recycled into soft-drink bottles, aluminum cans were recycled into aluminum cans, cardboard was recycled into cardboard, and so on. However, modern technology has

Sleeping on milk jugs? These pillows contain 58–90% recycled plastic.

made it possible for recovered materials to be purer and of a higher quality, which means more things can be made out of recycled materials. Scrap metals are used to make cans, nuts, bolts, screws, and tools. Some pillows are filled with a hypoallergenic polyester made from 58–90% plastic recycled from soft-drink bottles, milk jugs, and other plastics. Recycled plastics are also used to make toys and dolls.

Plastics and metals are not the only materials that can be recycled. Fabric manufacturers use recycled materials to make clothing. Leftover yarn and fabric are recycled into new yarn and fabric. Even the snippets that end up on the factory floor are recycled into threads. Fibers made from recycled soft-drink bottles are mixed with recycled cotton to make heavy fleeces or denim for blue jeans.

What will be made out of recycled products next? For the answer to that, just check the packages, clothing, and other products you buy. Tomorrow's fashions may be made from the soft-drink bottles you put in the recycling bin today.

Think About It

1. What is the difference between pre-consumer waste and post-consumer waste?

2. Which products in your house do you think are made from recycled materials? Check the labels of these products to find out.

3. Using the products you looked at in question 2, record the amount of pre- and post-consumer waste. Which products use the most pre-consumer waste? Which use the most post-consumer waste? Why do you think this is so? Consider what the products are made out of and where these materials might come from.

The Garbage Gazette

April 24 · Local Edition · Vol. 1, Issue 11

Nature's Garbage Disposal

Aristotle called them "the intestines of the earth." Charles Darwin considered them "indispensable." Most other people just call them "icky." What are these revered, revolting creatures?

Earthworms, the slimy wigglers who fill the sidewalks with their squishy bodies—and the air with their wormy odor—after it rains. Like most people, you've probably never given earthworms a second thought, unless you needed bait for fishing. But as landfills close and alternatives are needed, earthworms are entering the limelight doing the task they were born to do.

During digestion, an earthworm ingests decaying organic materials, absorbs what it needs, and secretes the leftovers, which are called "castings," as waste. The castings contain important plant nutrients that are more soluble to plants than nutrients from non-worm sources. Studies have shown that plants fertilized with castings yielded better results in growth tests than plants fertilized with other plant-growth formulae. But what does this have to do with waste disposal?

Ask a vermicomposter, a person who uses earthworms, usually of the *Eisenia foetida* (the tiger worm) or *Lumbricus rubellus* (the red worm) species, to turn organic waste into fertilizer. Instead of throwing away uneaten food, vermicomposters feed it to their worms, who in turn make castings, which are harvested for use as fertilizer. This keeps some organic

Using worms like this large red worm is becoming a popular waste management strategy. (Photo courtesy of Blair's Bait Farm, Clover, South Carolina.)

materials out of landfills and reduces the need for chemical fertilizers.

Vermicomposting companies are popping up across the nation, but the process is simple enough to do in your house or even in your school. Students in many U.S. schools scrape their food waste into pails instead of garbage cans. The contents of the pails are fed to earthworms kept in special bins. Students harvest the castings, which are then used by the school or sold to the community. One school in California fed over 1,636 kg of organic waste to their earthworms during one school year, saving over $6,000 in dumpster fees. The students used the castings as fertilizer for their garden, which supplies vegetables for the cafeteria salad bar.

Think About It

1. What is a vermicomposter? Where do you think the name comes from?

2. A good source for learning about vermicomposting is the book Worms Eat My Garbage by Mary Appelhof. How could you find other sources of information about vermicomposting?

3. How much food waste does your family throw away each week? Devise a plan to find out.

4. How much food waste does your school throw away each week? Devise a plan to find out.

Lesson 7:
Disposal Methods

This section examines current methods of landfilling and the problems of hazardous waste disposal. To study landfills, students first make a simple prototype of a landfill and then draw up their own designs for a landfill in the activity "**Design Your Own Landfill**." Then they play "**Believe It Can Rot—Or Not**," a simulated game show in which students compare decomposition rates between items buried in a landfill and those left exposed to nature's elements. Landfill issues are explored further in "**The Bottom Line(r)**," in which students observe the process of leaching and conduct an experiment to compare the effectiveness of various materials used in landfill construction to contain leachate.

The topic of hazardous waste is covered in the student activity "**Household Hazardous Waste**," in which students examine the labels of various hazardous products and prepare management plans for hazardous materials in their own homes.

While sanitary landfills are common today, they are a relatively new method of solid waste disposal. Students take a look at "**A Timeline of Trash Disposal**" in *The Garbage Gazette* included with this lesson.

● Teacher Background for "Design Your Own Landfill" and "Believe It Can Rot—Or Not"

Burying Waste

The open dump was the precursor to the modern sanitary landfill. In open dumps, trash was dumped and often burned. These dumps were not only eyesores but also breeding grounds for flies, cockroaches, mosquitoes, rats, and mice. They also caused odor problems, water pollution from runoff, and air pollution from burning. Federal legislation was enacted to phase out all dumps by 1983, but this deadline was not met. The Resource Conservation and Recovery Act (RCRA) of 1976 provided the legal base for federal regulation of sanitary landfills and, in combination with the Toxic Substances Control Act (TSCA) of that same year, provided strict regulation on production and disposal of toxic materials and a cradle-to-grave tracking system for such materials.

In 1991, the EPA established new, stricter landfill regulations, which now serve as the basis for all state landfill regulations. Enviroene, part of the U.S. EPA's World Wide Web site, summarizes the regulations as follows:

- Location restrictions—Landfills cannot be located near ecologically valuable wetlands or in areas prone to natural disasters, such as flooding or seismic activity. They also cannot be located near airports because birds drawn to landfills can pose hazards to flying aircraft.

- Operating requirements—Landfills must refuse to accept regulated hazardous waste, apply a daily cover, control disease vector populations, monitor methane gas, restrict public access, control stormwater runoff, protect surface water from pollutants, and keep appropriate records.

- Groundwater monitoring—Landfills must be designed to ensure that groundwater meets drinking-water standards. All landfills must have monitoring wells to detect contamination and must take full corrective action if contamination occurs.

- Closure and Post-Closure Costs—When a landfill stops accepting waste, it must be covered to keep liquids away from the buried waste. After closure, the operator is responsible for the final covering, leachate control, and monitoring of groundwater and methane gas for 30 years.

Two types of modern landfills exist: sanitary and secure. Strict guidelines regulate what constitutes sanitary and secure landfills. Sanitary landfills have liners to prevent contamination of soil, leachate monitoring systems to prevent the contamination of groundwater, and methods of collecting methane gas to prevent explosive pockets from building up in the landfill. (See Figure 1.)

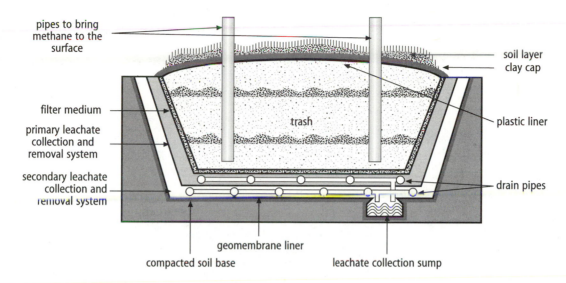

Figure 1: A sanitary landfill

Sanitary landfills are much safer than the open dumps of the past from the perspective that they are better insulated from the environment, but this too can have its disadvantages. The decomposition that occurs within landfills is dependent upon the activity of microorganisms. The design of landfills to minimize uptake of oxygen and water affects the type of bacterial activity and thus the decomposition process. The first stage of this process is some aerobic decomposition of the waste, primarily producing carbon dioxide, water, and nitrates. When the oxygen supply is depleted, the bacterial activity changes. Now other kinds of microorganisms, including anaerobic microorganisms, produce primarily volatile acids and carbon dioxide, resulting in an increase in acidity to a pH of 4–5. Later, methane-producing bacteria begin to predominate, converting the volatile acids to methane and carbon dioxide and raising the pH to near neutral values (7–8). The rate of decomposition and time required for each of these phases depends on a variety of factors, including moisture, temperature, rainfall, and permeability of the soil cover. Besides the slowed process of decomposition, sanitary landfills are relatively benign to the environment as long as the seals and monitoring systems are intact and the methane is released periodically. Some landfills collect the methane gas from decomposition and use it as a fuel source for the neighboring community.

Secure landfills are authorized to accept toxic waste and have much stricter safety precautions than sanitary landfills. Different types of toxic waste are entombed in separate chambers, and a careful inventory is kept of what items are buried there. (See Figure 2.)

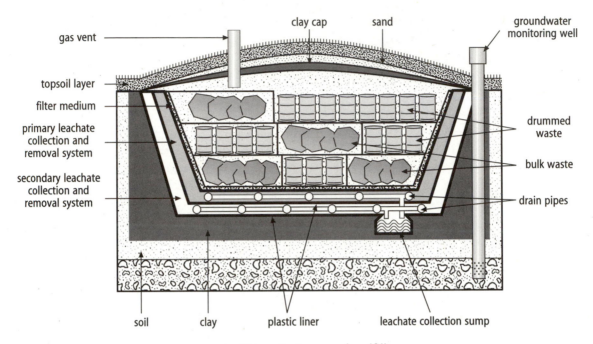

Figure 2: A secure landfill

Secure landfills require the highest degree of care in their design and upkeep. They are designed for the indefinite storage of toxic chemicals, and great care is taken to prevent leaks. However, if contamination of the surrounding environment were to occur, it would potentially be much more serious than contamination from a sanitary or unsanitary landfill because of the nature and concentration of the chemicals leaking out.

Reducing, reusing, recycling, composting, and incinerating cannot completely eliminate the need for landfills no matter how successfully these processes are implemented. Landfills will always be necessary if the municipal waste stream continues to have the same constituents as it does currently.

● Teacher Notes for "Design Your Own Landfill"

Students discuss potential environmental problems associated with burying solid waste in the ground, acquire a knowledge of landfill construction regulations in the Ohio Administrative Code, and apply this knowledge to design of landfills.

Group Size .... 5 students
Time Required .... 2–4 class periods

Materials

For the Opening Strategy
- plastic wading pool
- 2 pieces of PVC pipe long enough to lay across the bottom of the pool (approximately 1 m each)
- approximately 30-cm length of PVC pipe
- stopper or brace to hold the PVC tubing upright
- 2, 1.25-m x 1.25-m pieces of brown fabric to represent dirt
- 1.25-m x 1.25-m piece of green fabric or artificial turf to represent grass
- miniature playground equipment (such as from a Fisher Price™ play set)

For the Procedure
- paper
- pencils, markers, or crayons

Opening Strategy

1. Discuss with students the variety of materials and substances that are found in solid waste. Discuss the potential for land disposal of solid waste to generate leachate and methane and how these can become hazardous to the environment and to human health if not controlled at the landfill site. You may use the Common Uses of Hazardous Chemicals section of these Teacher Notes to consider the potentially hazardous chemicals that can be found in leachate and the products they can come from.

2. As a class, construct a mock landfill. Ask for seven volunteers to help with construction. These students will be the "landfill engineers."

3. Ask the class what factors need to be considered when planning a location for a sanitary landfill. *Avoid close proximity to water; avoid close proximity to large communities, to which stench and pests could present a nuisance; choose solid*

ground underneath, preferably with a layer of clay in the soil; provide easy access for large trucks and consider transportation costs, etc. Next ask the class why they think a rubber lining or a layer of clay would need to go in the bottom. *To make sure liquids from the landfill do not seep into groundwater, which some people may depend on for drinking water.* Have the student place the pit lining (the plastic pool) in the front of the classroom. Does the class think the landfill is now ready to accept trash? If they say no, ask what else it needs. If they say yes, ask questions about what happens when it rains or snows to develop the idea that liquids pooling up in the bottom of the landfill could be a problem.

4. Ask students what could be done to address the problem of pooling liquids. If necessary, lead the class to suggest that pipes might be used to help solve the problem. Tell the class that pipes are placed in the bottom of the pit before the garbage goes in. These pipes are used to collect liquids coming out of the landfill which might pool up on the liner, to test those liquids, and to prevent them from seeping into local water supplies. Have the second volunteer lay the two long pieces of PVC pipe in the bottom of the pit. Now does the class feel the landfill is ready to accept trash? If students say no, ask what else it needs. If they say yes, ask questions to get students to consider that methane gas buildup could be a problem but also a resource.

5. Ask the class what could be done to address the problem of decomposition gases building up in the landfill. Lead them to suggest that pipes could help collect the gas and that the gas could be used as a fuel source. Tell them that engineers run pipes vertically to the bottom of the pit. These pipes allow methane gas, a potential explosion hazard, to escape or be collected and used as a fuel. Have the third volunteer set the short piece of PVC pipe upright near the rim of the plastic pool and brace it to keep it upright.

The reason for placing the pipe near the rim of the pool is to make sure that its upper end will remain uncovered when the landfill is "capped" with the cloth. If you wish, you can cut a hole in the brown cloth "daily cover," the brown cloth "cap," and the green cloth "vegetative cover" to accommodate the vertical pipe in the center of the pool.

6. Can the class think of any other things they might need to install before the sanitary landfill is ready to accept trash?

7. Have another student put a layer of dirt (1.25-m x 1.25-m brown cloth) over the items in the landfill. Ask the students why they think it would be necessary to continue to cover trash items with dirt overnight and uncover them in the morning. *Overnight covering is necessary to prevent the spread of odors, keep birds and animals out, and prevent lightweight items from blowing away. The soil must be removed in the morning so that gases will not build up and be forced to move laterally.*

8. At this point, close the landfill. Have the fifth volunteer put a soil or clay layer (the other piece of brown cloth) on top of the landfill. Have the sixth volunteer place grass (the piece of green cloth) on top of the soil or clay layer. Ask students why they think grass is planted on top of landfills and why trees are not. *Grass roots hold the soil in place, but tree roots are so deep that they would penetrate to the trash and expose it.* Ask the class if they can think of any uses for this land now that it is covered up with soil and grass. If students suggest buildings, mention to them that buildings will cause the same kinds of problems as trees because of the basements. Also, buildings would be unstable because landfills settle with time. Have the last volunteer construct the playground on top of the landfill and discuss this as a possible use of the land.

 Save this mock landfill for use in the next activity, the game show "Believe It Can Rot—Or Not."

9. Tell the students that the model landfill they just completed is a very simplified version of a real landfill and that in this activity, they will draw up a more detailed design for a landfill, based on EPA regulations. Distribute the Student Instructions and Student Information and have students follow the Procedure.

Discussion

1. Show the class the diagram of a model landfill (provided in the Teacher Background for this activity). Discuss how students' designs compare to the drawing.

2. Review the Permit-to-Install Procedures section of the Student Information. Discuss the purpose and importance of each step in the Procedure.

3. Show students an overhead of the Sanitary Landfill Development Cost Estimate section of these Teacher Notes. This sheet includes phases of the construction and management process that students did not have to deal with in this activity and that they may not have thought about, especially Site Characterization and Post-Closure costs. Do any of the list items or costs surprise students?

Resource

"Designer Landfill"; *Investigating Solid Waste Issues;* Ohio Department of Natural Resources: Columbus, OH, 1996, pp D-11–D-21.

Common Uses of Hazardous Chemicals

The elements, compounds, and ions listed below are potential sources of groundwater, surface water, and soil pollution. Landfill operators must monitor groundwater for temperature, pH, and chemical species including the following:

Names	Common Uses
Acetone	Paint, varnish, nail polish remover
Acrolein	Pharmaceutical, herbicide
Acrylonitrile	Dyes, grain fumigant
Ammonia	Cleaning solutions, floor finish
Arsenic ion	Insecticides, preserving of hides
Barium ion	Ceramics, photographic paper
Benzene	Pesticides, medicine, insecticides
Bromodichloromethane	Solvent, fire extinguisher fluid
Bromoform	Medicine, solvent, fire inhibitor
2-Butanone	Paint removers, cements, adhesives
Cadmium, Cadmium ion	Batteries, pigments
Calcium ion	Used in manufacturing
Carbon Disulfide	Grain fumigants, soil disinfectant
Carbon Tetrachloride	Aerosols, propellants, spot removers
Chlorobenzene	Insecticide, pesticide
Chloroethane	Anesthetic, refrigeration
2-Chloroethyl Vinyl Ether	Anesthetic, sedative
Chloroform	Soil fumigant, insecticide
Chromium ions	Ceramic glazes, color additives
Copper	Ornamental materials, wiring
Cyanide ion	Used in extracting gold and silver
Dichlorodifluoromethane	Refrigerant, aerosol propellant
1,1-Dichloroethane	Paint, varnish removers, solvent
1,2-Dichloroethane	Paint, varnish removers, fumigant
1,1-Dichloroethylene	Synthetic fibers, adhesives
Trans-1,2-Dichloroethylene	Camphor, perfumes, refrigerant
Cis-1,3-Dichloropropylene	Soil fumigant
Trans-1,3-Dichloropropylene	Soil fumigant
Ethylbenzene	Solvent, production of styrene
2-Hexanone	Solvent for paint, varnishes
Iron	Wire, steel products
Lead, Lead ions	Lead-acid batteries, sealants, oils, solder
Manganese, Manganese ion	Steel products, cleaner for metals
Magnesium	Used in production of alloys
Mercury, Mercury ions	Paints, barometers, thermometers
Methylene Chloride	Herbicide, fumigant, coolant
Methyl Bromide	Fumigant, refrigerant
Methyl Chloride	Fumigant, degreasing agent, solvent
4-Methyl-2-Pentanone	Solvent for paints and varnishes
Nickel, Nickel ions	Plating, batteries, coins
Phenols	Disinfectants, dyes, epoxy resins, herbicides, pesticides
Phosphorus ion	Fertilizer, cleaning solution
Potassium ion	Fertilizer
Selenium ion	Glass, ceramics, rubber
Silver ion	Dental supplies, jewelry, tableware
Sodium ion	Anti-knock agent for engines
Styrene	Synthetic rubber, insulators
Sulfate	Rubber, paints, resins, dyes
Toluene	Medicines, adhesive solvent, perfume
1,1,2,2-Tetrachloroethane	Insecticide, paint, varnish
1,1,1-Trichloroethane	Aerosol propellants, pesticides
1,1,2-Trichloroethane	Solvent for fats, oils, and resins
Trichloroethylene	Anesthetic, medicine, refrigerant
Trichlorofluoromethane	Refrigerant, solvent, fire retardant
Vinyl Acetate	Latex paint, adhesives, textiles
M-Xylene	Solvent, aviation fuel, insecticides
O-Xylene	Motor fuels, pharmaceutical, dyes
P-Xylene	Vitamins, insecticides
Zinc, Zinc ion	Component of brass, galvanizing

The Sanitary Landfill Development Cost Estimate

The following costs are based on 1988 dollars for a 100-acre site with a 20-year working life and a 30-year closure period.

Site Characterization Costs

Feasibility Study	$100,000
Ecological Study	45,000
Archeological Study	30,000
Total Cost for Site Characterization	**$175,000**

Primary Development Costs

Land Acquisition	$1,500,000
Hydrologic/Geotechnical Studies	325,000
Civil Engineering Design and Site Engineering Support	
• Design	195,000
• Permits and Licenses	100,000
• Technical Support and Consultation	125,000
Legal Consultation	360,000
Total Cost for Preliminary Development	**$2,605,000**

Final Development Costs

Clearing and Grubbing	$300,000
Excavation and Stockpile	6,450,000
Liner and Leachate Collection	
System Installation (1 of the following)	
• Single Natural Liner	14,200,000
• Single Composite Liner	21,500,000
• Double Composite Liner	42,500,000
Leachate Management	
• Pumps and Pipe Installation	450,000
• Leachate Pretreatment Facility	4,000,000
Surface Water Controls	
• Sedimentation Pond Construction	27,000
• Ditch Construction	54,450
Final Cover Construction (1 of the following)	
• Natural Clay Cover	5,700,000
• Synthetic Liner	9,800,000
Gas Management System	
• System Design	70,000
• Gas Monitoring Program	5,000
• Gas Migration Assessment	100,000
• Installation	1,630,000
Groundwater Monitoring System	
• Well Installation	200,000
Site Structures	
• Maintenance Building/Offices	2,000,000
• Weighing Scale	50,000
• Scale House	50,000
• Fencing	140,000
Total Cost for Final Development (Assume a Single Composite Liner with a Natural Clay Cover)	**$42,726,450**

Environmental Management Costs

Leachate Management	
• Treatment and Truck Transport	$4,380,000
Gas Monitoring and Control	
• Maintain Gas Probes	48,000
• Monitor Gas Probes	384,000
• Calibrate Equipment	44,000
Groundwater Monitoring	
• Well Maintenance	260,000
• Maintain Equipment	4,000
• Ground/Surface Water Testing and Analysis	1,920,000
Total Cost for Environmental Management	**$7,040,000**

Post-Closure Costs

Inspection	$240,000
Land Surface Maintenance	3,300,000
Leachate Management System	7,520,000
Gas Management	480,000
Groundwater Monitoring	1,864,000
Total Cost for Post-Closure	**$13,404,000**

Grand Total	**$65,950,450**

Landfill Capacity in the U.S., National Solid Waste Management Association, Washington, DC.

● Student Instructions for "Design Your Own Landfill"

In this activity, you will discuss potential problems associated with burying solid waste in the ground (landfilling), learn about landfill regulations, and use this knowledge to design a landfill.

Procedure

1. Consider various environmental problems that may arise when burying solid waste in the ground by completing the Potential Problems and Solutions section of this handout. Based on this list, think of some possible technological methods that can be applied to prevent these problems from occurring.

2. Use the Design Specifications for a Sanitary Landfill section of the Student Information to compare your list of Applied Technology with actual requirements.

3. Work in groups to design a model landfill by incorporating the design elements described in the handout. Each student in the group can be responsible for designing a specific system within the landfill. Examples of the specific design components are as follows:
   - foundation;
   - compacted soil base;
   - geomembrane (plastic liner);
   - groundwater control system;
   - leachate management system;
   - explosive gas system;
   - cap system (including surface water control); and
   - operational procedures.

   Include a brief explanation of the function of each component with a rough sketch.

4. After each component has been planned, integrate the components (including explanations) into a design for the entire landfill. Present your design to the class.

Potential Problems and Solutions

Old-fashioned dumps were places where garbage was taken and dumped on the land. Older landfills were places where a hole was dug and refuse was compacted and covered with soil. List problems for the environment and for human health that can arise when garbage is dumped on land or buried in the ground. Assign a letter to each problem in the list ("a," "b," "c," etc.).

Potential Environmental and Health Problems

_____

_____

_____

_____

_____

_____

For each potential problem listed above, think of a way that it could be dealt with to protect the environment and human health. Briefly explain the technology and method you propose, matching each item ("a," "b," "c," etc.) from the list above.

Applied Technology

_____

_____

_____

_____

_____

_____

● Student Information for "Design Your Own Landfill"

Design Specifications and Operational Procedures for a Sanitary Landfill

According to the Ohio Administrative Code [OAC 3745-27-06, (C) (3)], landfill permit-to-install applications submitted to the Ohio EPA must detail the measures and operations to control and manage the following:

- leachate production and migration;
- groundwater infiltration;
- explosive gas migration, if necessary;
- fires, dust, scavenging, vectors (rats), erosion, blowing litter, and birds; and
- surface water run-on and run-off and sediment discharge.

A landfill that includes methods for dealing with these concerns is considered a sanitary landfill. The following two sections describe the design specifications and operational procedures that are required by law.

Design Specifications

New sanitary landfills must be constructed to protect the environment and human health by using the "best available technology" (BAT). The following list describes methods that include BAT for controlling each concern listed above. In order to obtain certification as sanitary, a landfill must include the following:

- foundation preparation;
- compacted soil base;
- geomembrane (plastic liner);
- leachate management system;
- permanent groundwater control structures, if required;
- explosive gas control/extraction system, if required; and
- cap system.

Each of these components must meet particular specifications for design and construction. The following lists are some of the criteria for each component. (These criteria have been simplified for classroom use.)

Foundation Preparation
The site must be prepared on a surface that shall

- be freed of debris or harmful material;
- be able to bear the weight of the landfill and its construction and operations without causing or allowing a failure of the liner to occur through settling; and
- not have any abrupt changes in steepness that may result in damage to liner material.

Compacted Soil Base

A compacted soil base

- must be constructed of soil with 100 percent of the particles having a maximum dimension smaller than 2 inches;
- will be placed at bottom and exterior excavated sides of the landfill; and
- will be at least 5 feet thick.

Geomembrane of Polymer Material

A geomembrane (plastic liner) must be placed on the compacted soil base. It must

- prevent all but negligible amounts of fluid from leaching out; and
- be physically and chemically resistant to chemical attack by the solid waste leachate or other materials which may come in contact with the geomembrane.

Leachate Management System

A leachate management system must be designed to prevent clogging and crushing of the system and to limit the level of leachate to a maximum of 1 foot. The leachate management system must consist of

- a drainage layer placed on top of the geomembrane that is able to rapidly collect leachate entering the system, consists of a granular material, and is at least 1 foot thick;
- a means of removing leachate from the bottom of the landfill, with collection pipes embedded in the drainage layer;
- a protective layer to protect the compacted soil base, geomembrane, and leachate collection system from the intrusion of objects during construction and operation;
- lift stations with pumps that shall automatically commence pumping before the leachate elevation reaches high levels; and
- a leachate collection system in which leachate shall be treated and disposed of on-site, be treated on-site and disposed of off-site, or be treated and disposed of off-site.

Groundwater Control System

This system is required if solid waste placement is within 1,000 feet of an inhabited structure and if the landfill accepts garbage that decomposes, especially garbage that produces foul odors as it decays. Any permanent groundwater control structures shall adequately control groundwater infiltration through the use of nonmechanical means such as impermeable barriers or permeable drainage structures.

Explosive Gas Control System

This system is required if solid waste placement is within 1,000 feet of an inhabited building and if the landfill accepts garbage that rots. Explosive gas control structures shall be designed so that explosive gas cannot travel sideways from the sanitary landfill facility or accumulate in occupied buildings. Explosive gas control/extraction systems shall be designed in such a manner as to prevent fires within the area where the solid waste is buried. Construction of the explosive gas control/extraction systems shall not damage or disturb the cap system, the leachate management system, or the compacted soil base.

Cap System

Construction of a cap system (installed when the landfill is closed) shall minimize infiltration and consist of

- a compacted soil barrier layer, a minimum of 2 feet thick, constructed like the soil base;
- a granular drainage layer on top of the soil barrier layer, constructed like the drainage layer of the leachate management system;
- a complete and dense vegetative layer, consisting of soil and plants (not trees), placed on top of the granular drainage layer; and
- sloping of all land surfaces to prevent pooling of water where solid waste has been placed.

Operational Procedures

- Solid waste admitted to the landfill must be spread in layers and compacted to the smallest volume practical.
- Solid waste shall be covered daily by the end of each working day, and in no event shall solid waste be exposed for more than 24 hours after unloading. In most cases, a 6-inch layer of soil is used to cover material. However, the cover should be removed daily to make sure gas does not move sideways.
- The sanitary landfill shall be properly sloped and provided with additional drainage facilities.
- All new sanitary landfill facilities, all expansions of older landfills, and all sanitary landfills being closed shall have a groundwater monitoring system installed at appropriate locations and depths, to yield groundwater samples.
- If a landfill (new, old, or closed) is located so that a residence or other occupied building is located within 1,000 feet horizontal distance from buried solid wastes, an explosive gas monitoring system must be installed that will monitor all explosive gas pathways representing a potential hazard.

Permit-to-Install Procedures (in Ohio)

Prior to building a solid waste landfill (or any other solid waste facility), the builder must submit a permit-to-install (PTI) application to the Ohio EPA for review. The following explains the sequence of this initial permitting process and of the continuing permitting process during operation of the landfill. (These steps have been simplified for classroom use.)

1. A preliminary site survey is conducted to determine the best site for building the landfill based on siting criteria established by the Ohio EPA.

2. PTI application is prepared and sent to the Ohio EPA. Upon receiving a PTI application, Ohio EPA determines whether the site is suitable for a landfill. If the location meets siting criteria, the engineering review continues. If the location does not meet siting criteria, the application is recommended for revisions or denial.

3. When approval is obtained, construction of the landfill begins. As each component of the landfill is being constructed, the builder prepares a certification report. The Ohio EPA generally inspects each stage of construction.

4. After all components have been certified, the Ohio EPA issues a written certificate for the landfill describing the long-term plan for construction and operation of the landfill. Upon obtaining the certificate, the landfill operator can begin accepting wastes in accordance with the provisions of the certification.

5. Each year, the operator of the landfill must obtain an annual solid waste operating permit from either the local health department (if it is on the Ohio EPA's approved list) or the Ohio EPA. The permit is issued after inspection by the health department or the Ohio EPA of the operators' records and of the landfill itself. Any problems are reported by the inspector.

● Teacher Notes for "Believe It Can Rot—Or Not"

Through this simulated game show, students compare natural decomposition rates to decomposition rates in a sanitary landfill.

Due to the nature of this activity, no Student Instructions are provided.

Group Size .... Class or school assembly
Time Required .... Getting Ready: 20 minutes
Procedure, Part A: 20 minutes
Procedure, Part B: 40 minutes

Materials

The more props you have, the more fun and realistic this game show will seem. The following is merely a list of suggested materials for props. With a little imagination, this game show can be customized to utilize available materials and "actors."

For Getting Ready
Per class
- 10 pieces of posterboard or construction paper

For the Procedure
Per class
- 10 question-and-answer signs (See Getting Ready.)
- wading-pool "landfill" constructed in the "Design Your Own Landfill" activity
- 2 or 3 large, tree-like plants or cardboard forest
- watering can
- flashlight
- clear jar containing soil
- 2–3 latex balloons
- costume for Mother Nature such as wig, dress, wreath for head, small stuffed animals, flowers, leaves, animal puppet
- costume for the Master of Ceremonies such as tuxedo, bow tie, cummerbund, microphone
- 2 of each of the following "trash" items
    ○ glass jars
    ○ disposable diapers
    ○ aluminum cans
    ○ bimetal cans
    ○ leather objects (Old shoes work well.)
    ○ plastic bags

- pieces of wood (blocks)
- socks
- newspapers
- banana peels
- 20 small prizes to award for correct answers during the game show
- (optional) buzzer and light

For the Extension
Per demonstration
- 5 mL 70% isopropyl rubbing alcohol
- 2-L plastic soft-drink bottle and cap
- wooden splint or craft stick
- tongs
- match or Aim 'n Flame®

Safety and Disposal

No special safety or disposal procedures are required for the Procedure.

If doing the Extension, proper fire safety should be exercised, such as working on a flame-resistant surface and removing unnecessary flammable materials from the area. Long-haired people should tie hair back when working near a flame. Wear goggles while preparing for and doing the Extension.

Getting Ready

1. Use the 10 pieces of posterboard or construction paper to make double-sided signs according to the table in the "Answers to Game Show Questions" section at the end of this activity. The front of each sign should contain the four possible responses, and the back should contain the correct answer for the item "left out in nature." (The correct answer for the landfilled item will always be "d.") Organize the finished cards to follow the script.

2. Set up the wading-pool landfill constructed in the "Design Your Own Landfill" activity at one end of the presentation area or stage.

Opening Strategy

The class will be the interactive audience for a game show called "Believe It Can Rot—Or Not." The "guest" is Mother Nature. The teacher and an adult helper should play the roles of the Master of Ceremonies (MC) and Mother Nature (MN).

Ask the audience the following questions: Where does your trash go? What is the difference between a landfill and a dump? Does anybody know what a sanitary landfill is?

Procedure

Part A: Constructing the Set

Construct the decomposition chamber—in this case a forest environment. Place the large plants or cardboard forest at the opposite end of the stage or presentation area from the landfill. Ask students if they have ever seen plants or animals decomposing on a compost pile, a forest floor, or the side of a road. What factors do they think influence the process of decomposition? Keep querying the class until they come up with at least the following factors: microorganisms, water, heat, and air. As each student suggests one of these factors, have them move an appropriate representative material into the "forest." For the representative materials, use a watering can (for water), a flashlight (for sunlight, which produces heat), a jar of soil (for microorganisms), and a balloon (for air).

You can blow up the balloon ahead of time or have a student do it when air is suggested as a factor in the process of decomposition.

Part B: The Game Show

1. Introduce the game show. The following is a suggested script:

MC: Studio audience, welcome to "Believe It Can Rot—or Not," the game show which challenges your decomposition knowledge. I'm your host, [your name], and I would like to introduce my lovely and talented guest, Mother Nature. Please welcome Mother Nature (claps). Thank you for joining us today, Mother Nature, and thank you for the beautiful day.

MN: It's a pleasure to be here, [name of MC].

MC: Okay, studio audience, are you ready to play "Believe It Can Rot—or Not"? All right then, let's have our first two contestants. (Pick two members from the class and insert their names for "A" and "B.") [A] and [B], come on down. You're our first two contestants. Now last night, the two of you were out collecting fireflies in these glass jars. You were both startled by a sudden loud noise, dropped your jars and broke them. [A], you are somewhat of a slob. You leave your broken glass on the ground right there in the forest where a person or an animal could step on it and become injured. (Have the student place the glass jar in the mock forest.) [B], you are a neat and tidy person, you carefully picked up your glass and put it in the trash can to go to the landfill. (Have the student place the glass jar in the mock landfill.) Now [A], how long do you think it will take for that glass jar to completely decompose and for the elements to return to nature when it's left out in the forest? (Show the student the options on the card and read them out loud. Wait for the student to select an answer.) Mother Nature, is [A] right that it will take [answer] for a glass jar to decompose completely out in the forest?

For each scenario, Mother Nature tells both contestants whether or not they are right. The time needed for items buried in the landfill to completely decompose will always be "d) none of the above."

MN: (Hit the light or the buzzer if available; if not just reply "No, I'm sorry" or "Yes, that's right.") I estimate it takes 1 million years for a glass jar to decompose if left out to my elements. (Give [A] a prize or just thank the student for playing.)

MC: Now [B], your glass jar was buried in a landfill. How long do you think it will take for it to completely decompose? (Again, show the student the options on the card, read them aloud, and wait for the student to select an answer.) Mother Nature, is [B] right that it will take [answer] for the glass jar to decompose in the landfill?

MN: (Hit the light or the buzzer if available; if not, just reply "No, I'm sorry" or "Yes, that's right.") The answer is none of the above, I estimate that a glass jar simply will not decompose if buried in a landfill. (Give [B] a prize or just thank him/her for playing.)

2. Continue the game show with a similar script for each item. Have two new contestants come up to the stage for each item, with the students putting their items either in the mock landfill or in the mock forest. The item scenarios could be as follows:

- Disposable diaper: You were out on a walk with your baby sisters, and you had to change their messy diapers. [A] was grossed out by the smell and left one in the woods. [B] wrapped the other in a plastic bag, put it in the diaper bag and threw it away when [he/she] got home.

- Aluminum can: You were on a picnic in the woods and squirrels ran up and stole your aluminum soft-drink cans. [A] decided to let them have the can and went home. [B] decided to follow them until they put the can down so [he/she] could dispose of it properly.

- Bimetal can: You were walking along the creek and found a couple of old bimetal cans that some picnickers had left. You decided to use them to dump creek water over each other's heads. When you were done [A] didn't think the bimetal can was [his/her] responsibility and threw it back on the ground where [he/she] found it. [B] decided it was better to pick up litter that wasn't [his/hers] than to leave it and took it to the nearest trash can.

- Leather: You were hiking and your leather shoes got so gunked up with muddy creek water that [A] decided just to leave them in the woods. [B] took them home and threw them away.

- Plastic bag: You were collecting leaves for botany class in a plastic bag. The wind picked up and blew the bags out of your hands up into a tree. [A] decided it was too much bother to try and get [his/her] bag and began gathering leaves all over again. [B] went through the effort of retrieving the other bag, which had a hole in it, and threw it out when [he/she] got home.

- Wood: You were sitting on a picnic table in the woods, and you broke it. [A] ran away from the scene of the crime. [B] took the broken piece to the ranger's office to confess. The ranger said she would see that the table got repaired and threw the broken piece away.

- Sock: You were hiking so long and so hard that you wore holes through the toes of your socks. Since you both had brought spare socks with you on the camping trip, [A] just left the damaged pair in a hollow log. [B] packed them up to dispose of when [he/she] got home.

- Newspaper: It was a nice day so you decided to take the newspaper outside and enjoy the sunshine while you caught up on world events. The wind picked up and blew the newspaper out of your hands. [A] decided to go back inside. [B] picked up all the pieces of the newspaper, which were now wet and muddy, and put them in the trash can.

- Banana peel: You decided to have a little snack while you were hiking. [A] just tossed the banana peel on the ground. [B] held onto it until [he/she] passed the next trash can.

3. Ask the students to explain why the items decompose in the forest but not in the landfill. Review the decomposition factors present in the forest. Ask which, if any, of these factors are present in landfills and whether the supply is continuous or limited. Ask students whether all items that were sent to the landfill really needed to go there. What items could have been composted, reused, recycled, or incinerated?

4. Discuss how the construction of a landfill differs from a compost pile (see Lesson 6) and what factors influence how readily materials decompose in each.

5. Discuss the concept of littering. Do students think that, because items left outside will decompose faster than items buried in a landfill, littering is a good way to dispose of all waste? Discuss reasons why it may be better to dispose of the materials mentioned in the game show in a landfill than to leave them in the forest.

Extension

1. The need for methane recovery systems can be demonstrated as follows:

   The teacher or another adult demonstrator must wear goggles and use caution during this demonstration. Do not tell students what flammable gas is in the bottle beyond telling them it is similar to methane. This could invite home experimentation with flammable liquids. Larger bottles or long tubes can be used instead of the 2-L soft-drink bottle in this experiment. Practice first to see how much rubbing alcohol is needed and what kind of fireball results.

   a. A few moments before students arrive, pour one capful (approximately 5 mL) 70% isopropyl rubbing alcohol into a plastic 2-L soft-drink bottle. Put the cap on the bottle and swirl the bottle so that the liquid coats the sides all the way to the top.

   b. Tell students a gas similar to methane is trapped in the bottle. Explain that methane gas is formed during anaerobic (in the absence of oxygen) decomposition, which takes place to some extent inside landfills. This gas can build up in large pockets within a sealed landfill unless some type of removal system is installed. The capped bottle is like a pocket of methane gas trapped inside a landfill.

   c. Remove the cap. Ask students what is in the bottle now. *Opening the bottle allowed air to enter, introducing oxygen to the system.* (Explain that in this demonstration, oxygen was already present in the bottle before the alcohol was added, but in the system it represents, there would not have been.) Set the bottle on a table or other smooth surface. Holding a wooden splint or craft stick in a pair of tongs, use a match or an Aim 'n Flame to light it. Being careful to keep hands away from the opening of the container, lower the ignited end of the splint in. The alcohol gas will ignite, flames may shoot out the top, and the container will briefly contain a fireball.

   d. Ask students why it would be dangerous to leave pockets of methane in a landfill. What conditions would be needed to ignite the methane? *Oxygen and a spark, such as from lightning. If methane has seeped into the basement of a nearby building, electricity in the building could ignite it.* Can the class think of anything useful that could be done with the methane gas once it is collected? *It could be sold as is or used to generate electricity.*

Cross-Curricular Integration

Language arts:

- This activity can also be carried out with the students as the principle actors instead of the teachers and other adults. Have a small group of students write their own script and carry out the play for a younger grade level.

Social studies:

- Discuss the different factors that affect the location of landfills and incinerators. Should landfills be close to communities or far away? Weigh the economic benefits of low trucking costs against the increased health risk of having a landfill close to communities where drinking water, soil, and air could become contaminated.

Explanation

Modern landfills fall into two categories: sanitary, the type designed for municipal solid waste; and secure, designed for toxic chemical disposal. In order to be considered sanitary, a landfill needs the following components:

- foundation preparation;
- compacted soil base;
- geomembrane (plastic liner);
- leachate management system;
- permanent groundwater control structures, if required;
- explosive gas control/extraction system, if required; and
- cap system.

These components work together to protect the soil, groundwater, surface water, and air from potentially hazardous components in the landfill.

Because such great care is taken to keep materials from escaping the landfill into the surrounding environment, air and water from the surrounding environment have a difficult time getting into the landfill. These components are necessary for aerobic decomposition to take place. Without them, aerobic decomposition is slowed and eventually stopped, anaerobic decomposition takes over, and methane gas is produced.

Methane is a combustible gas. When exposed to oxygen and a spark, it can ignite and explode. Methane gas unmonitored and uncollected can gather in the basements of houses near the landfill or form explosive pockets near the surface of the landfill. A spark from metals rubbing together or lightning could cause it to ignite if oxygen is available.

Answers to Game Show Questions

	Side 1: Questions		Side 2: Answers (for "left out in nature")
Glass jar:	a) 1,000 years b) 10,000 years	c) 1 million years d) none of the above	Glass jar: c) 1 million years
Disposable diaper:	a) 3–5 years b) 30–50 years	c) 300–500 years d) none of the above	Disposable diaper: c) 300–500 years
Aluminum can:	a) 2–5 years b) 20–50 years	c) 200–500 years d) none of the above	Aluminum can: c) 200–500 years
Bimetal can:	a) 10 years b) 100 years	c) 1,000 years d) none of the above	Bimetal can: b) 100 years
Leather:	a) 5 years b) 50 years	c) 500 years d) none of the above	Leather: b) 50 years
Plastic bag:	a) 2–3 years b) 20–30 years	c) 200–300 years d) none of the above	Plastic bag: c) 200–300 years
Wood:	a) 2 months b) 2 years	c) 20 years d) none of the above	Wood: c) 20 years
Sock:	a) 1–5 weeks b) 1–5 months	c) 1–5 years d) none of the above	Sock: c) 1–5 years
Newspaper:	a) less than 1 week b) 1 week–1 month	c) 1 month–1 year d) none of the above	Newspaper: c) 1 month–1 year
Banana peel:	a) 2–5 days b) 2–5 weeks	c) 2–5 months d) none of the above	Banana peel: b) 2–5 weeks

● Teacher Notes for "The Bottom Line(r)"

In this activity, students will observe the process of leaching and conduct an experiment to compare the effectiveness of various liner materials used in landfill construction at containing leachate.

Group Size .... 5 students
Time Required .... 75–80 minutes for Getting Ready
15 minutes for the Opening Strategy
30–40 minutes for the Procedure

Materials

For Getting Ready
Per demonstration
- 6 film canisters with lids
- permanent marker
- 6 labels or masking tape
- liquid laundry detergent with optical brightener such as liquid Tide®

Per class
- sharp knife
- scissors
- pushpin or thumbtack
- Leachate Measurement Scale (provided at the end of these Teacher Notes)
- (optional) cheese grater

Per group
- 10, 2-L plastic soft-drink bottles

Have your students help you collect 2-L bottles in advance of this activity.

For the Opening Strategy
Per class
- colander
- mixture of common trash items such as the following:
  ◦ paper
  ◦ aluminum foil
  ◦ plastic bottle caps
  ◦ craft sticks
  ◦ glass marbles
- 6 film canisters prepared in Getting Ready

- watering can with water
- clear bowl or container to catch water
- black light

For the Procedure

Per group
- 5 leachate collection apparatuses (prepared in Getting Ready)
- ruler marked in inches
- sand
- aquarium gravel
- potting soil
- water-based modeling clay
- plastic garbage bag
- tape
- 1-L beaker
- water
- watch or clock

Safety and Disposal

Use caution when cutting the 2-L bottles. The detergent and water from the demonstration can be poured down the drain.

Getting Ready

For the Opening Strategy

1. Using the permanent marker and labels, label the six film canisters as follows: oil, oven cleaner, bug spray, antifreeze, bleach, and paint.

2. Half-fill each of the canisters with the laundry detergent and replace the caps loosely so that they will fall off when the canisters are tipped.

For the Procedure

1. Cut 10, 2-L soft-drink bottles in half by starting the cuts with a sharp knife and then using scissors to cut the rest of the way. Set aside the top halves. Students will use only the bottom halves for this activity.

2. Using a pushpin or thumbtack, poke a hole in the tip of each bulge at the bottoms of five of the bottle halves. These will be the testing containers.

3. Tape a Leachate Measurement Scale (provided) to the side of each remaining unperforated bottle half. The bottom of each scale should start at the tip of one of the bulges of the base of the bottle and extend straight up along the side. (See Figure 1.)

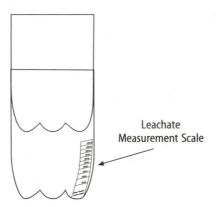

Leachate
Measurement Scale

Figure 1: Tape the Leachate Measurement Scale to the bottle.

4. Place each of the five perforated bottle halves inside one of the unperforated bottle halves. These stacked bottles are the leachate collection apparatuses.

5. (optional) If the water-based modeling clay is too hard to be molded into the 2-L bottles easily, you may want to use a cheese grater to grate the clay. These small pieces can be packed much more easily into the bottles.

Opening Strategy

1. Ask the class if they can think of any items that may end up in landfills and be potentially harmful to humans. Besides the items they can see (broken glass, oil, etc.), are there any items they might not be able to see which could do harm? Tell students that you are going to demonstrate a type of pollution they may not have thought about before. Demonstrate this pollution as follows:

   a. Hold up the colander and explain to the class that it represents a landfill. Ask students what types of trash items (in terms of material, such as paper, metal, plastic, wood, and glass) typically go to the landfill. As the students name the different types of items, hold the colander above the catch basin and put representative pieces of trash into it. If they do not mention toxic items, bring up the category and read the labels on the film canisters while adding them to the colander. Be sure that the canisters fall over and the laundry detergent spills over the other trash items (at this point it is important that the colander be held above the catch basin in case it starts dripping).

b. Simulate rain falling on a landfill by sprinkling water over the colander with the watering can. Explain to students that until a landfill is capped, a significant amount of rainwater or melted snow seeps (percolates) through the trash, picking up chemicals from the trash on its way. This process is called "leaching," and the contaminated liquid is called "leachate." If a leachate collection system is not intact, this water can enter the community water supply, which could be groundwater and/or bodies of water.

c. Ask students how many ways the community water supply can be used. *Drinking water, wildlife, recreation, fishing.*

d. Show students the water supply in the catch basin and ask them if it appears to be clean. Discuss the importance of a healthy water supply to a community. Ask students if they feel that the water supply in this demonstration community is healthy. How do they know whether it is or not?

e. Turn out the lights and shine a black light at the water in the catch basin. "Contaminated" water will glow under the black light. Explain to students that clear, colorless water is not always healthy water.

2. Explain to the students that modern landfills are designed to prevent or minimize water pollution. Review the parts of the landfill, which students should recall from the first activity of this lesson, "Design Your Own Landfill." Leaching is controlled by both pumping and the materials used in construction. Review the different layers in the closed, capped landfill, especially the different kinds of soils used.

3. Distribute the materials and have students follow the Procedure in the Student Instructions sheet. After the students have finished the Procedure, discuss how the results students observed apply to landfill construction. Have students consider the purpose of each layer.

Extension

Use the demonstration in the Opening Strategy as an introduction to the concept of bioaccumulation. Explain that if the "toxic chemical" located in the water supply was not water-soluble and was, therefore, readily stored in animal tissue, the concentration would increase in those animals higher up in the food chain. For instance, if a human ate an average of three contaminated fish a day, the amount of the toxic chemical in all of those fish could stay in the human even though the water-soluble nutrients would be excreted. The concentration of the toxic chemical would continue to get higher and higher even though the total volume (the human) would stay relatively constant. Eventually, toxic chemicals can build up to such high levels that they prove fatal.

Cross-Curricular Integration

Earth science:

• Have students study the concept of soil profiles and learn about typical soil profiles of various geographic regions. Soil maps for your town or county may be available from your local soil and water conservation district. Discuss the suitability or unsuitability of various soil types for landfill construction.

Life science:

• The demonstration in the Opening Strategy reinforces the concept that water can look clean but really be unsafe to drink. Discuss how important it is that community water supplies go through a water treatment plant where contaminating chemicals and pathogens can be removed. Residue from unclean water can remain on foods and other surfaces, so it is important to wash foods and hands with clean water before eating.

Explanation

Many wastes are potentially hazardous to humans, animals, and the environment and may need special treatment. Unfortunately, some of those wastes can make it into landfills that were not designed to accommodate their toxic nature. In these landfills, water can percolate through, taking hazardous substances with it. This process is called leaching. Even though it may be clear and colorless, this leachate can cause environmental damage and endanger the life forms it encounters. Properly constructed landfills can prevent or minimize the escape of leachate through the use of liners, different kinds of soil, and pumping.

This activity focuses on the use of different construction materials to prevent the escape of leachate and aid in its collection. The percolation rate of soil determines how quickly or slowly water and other substances move through a sample of soil. This rate also influences how quickly or slowly a hazardous substance might pass through the soil to reach groundwater. Soil with a slower percolation rate will hold water, and hazardous substances will pass through it more slowly. The percolation rate can be determined by how much water passes through the soil after a given amount of time.

Resource

"Conducting Percolation Rate Experiment"; *Investigating Solid Waste Issues;* Ohio Department of Natural Resources: Columbus, OH, 1996, pp B-99–B-100.

Leachate Measurement Scales

500 mL	500 mL	500 mL	500 mL	500 mL
450 mL	450 mL	450 mL	450 mL	450 mL
400 mL	400 mL	400 mL	400 mL	400 mL
350 mL	350 mL	350 mL	350 mL	350 mL
300 mL	300 mL	300 mL	300 mL	300 mL
250 mL	250 mL	250 mL	250 mL	250 mL
200 mL	200 mL	200 mL	200 mL	200 mL
150 mL	150 mL	150 mL	150 mL	150 mL
100 mL	100 mL	100 mL	100 mL	100 mL
50 mL	50 mL	50 mL	50 mL	50 mL
where "beaker" meets counter ↓	where "beaker" meets counter ↓	where "beaker" meets counter ↓	where "beaker" meets counter ↓	where "beaker" meets counter ↓

500 mL	500 mL	500 mL	500 mL	500 mL
450 mL	450 mL	450 mL	450 mL	450 mL
400 mL	400 mL	400 mL	400 mL	400 mL
350 mL	350 mL	350 mL	350 mL	350 mL
300 mL	300 mL	300 mL	300 mL	300 mL
250 mL	250 mL	250 mL	250 mL	250 mL
200 mL	200 mL	200 mL	200 mL	200 mL
150 mL	150 mL	150 mL	150 mL	150 mL
100 mL	100 mL	100 mL	100 mL	100 mL
50 mL	50 mL	50 mL	50 mL	50 mL
where "beaker" meets counter ↓	where "beaker" meets counter ↓	where "beaker" meets counter ↓	where "beaker" meets counter ↓	where "beaker" meets counter ↓

● Student Instructions for "The Bottom Line(r)"

In this activity, you will observe the process of leaching and conduct an experiment to compare the effectiveness of various materials used in landfill construction at containing leachate.

Procedure

1. Place materials in the leachate collection apparatuses as follows (use a ruler to ensure consistency between the depths of the layers:
   a. Container #1—2 inches of sand
   b. Container #2—2 inches of gravel
   c. Container #3—2 inches of potting soil
   d. Container #4—2 inches of clay
   e. Container #5—layer of plastic from a garbage bag, taped to the inside

2. For each apparatus, measure 500 mL water in the large beaker. Pour this water on top of the material in each leachate collection apparatus. Record the starting time (the time at which you poured the water in the apparatus) for each container on the Leachate Collection table in the Data Recording section.

3. At 5-minute intervals (up to and including 20 minutes), read the scale on the bottom half of the apparatus to determine the amount of leachate that has passed through each test container into its collecting container. Record this amount on the Leachate Collection table.

4. Answer questions a–c in the Analysis Questions section of this handout.

Data Recording

Table: Leachate Collection

Experiment # 1	Starting Time	Volume (mL) in 5 mins	Volume (mL) in 10 mins	Volume (mL) in 15 mins	Volume (mL) in 20 mins
Container 1					
Container 2					
Container 3					
Container 4					
Container 5					

Analysis Questions

a. In which container did water pass through the fastest?

_____

_____

_____

b. In which container did water pass through the slowest (or not at all)?

_____

_____

_____

c. In what order (from top layer to bottom layer) might these materials be found in a real landfill?

_____

_____

_____

_____

_____

● Teacher Background for "Household Hazardous Waste"

What Are Hazardous Substances?

A hazardous chemical is a chemical that poses some form of danger to humans or the environment. More specifically, according to the Resource Conservation and Recovery Act of 1976 (RCRA), materials are considered hazardous if they "cause or significantly contribute to an increase in mortality or an increase in serious irreversible, or incapacitating reversible illness; or pose a substantial present or potential hazard to human health or the environment when improperly treated, stored, transported, disposed of, or otherwise managed." Almost every home in the United States contains hazardous chemicals of various kinds. Hazards posed by chemicals can be classified into seven basic types: flammable/combustible, explosive/reactive, sensitizer, corrosive, irritant, carcinogen, and toxic (poisonous). Many hazardous chemicals may fall into more than one of these categories. Descriptions of these categories are as follows:

- Flammable/Combustible: Describes any material that ignites easily and burns rapidly.

- Explosive/Reactive: An explosive material produces a sudden, almost instantaneous release of pressure, gas, and heat when subjected to abrupt shock, high temperature, or an ignition source. A reactive substance will vigorously undergo a chemical change under conditions of shock, pressure, or temperature.

- Sensitizer: A material that on first exposure causes little or no reaction in humans or test animals, but upon repeated exposure may cause a marked response not necessarily limited to the contact site. Skin sensitization is the most common form. Respiratory sensitization to a few chemicals also occurs.

- Corrosive: A chemical that causes visible destruction of or irreversible alterations in living tissue by chemical action at the site of contact.

- Irritant: A noncorrosive material that causes a reversible inflammatory effect on living tissue by chemical action at the site of contact as a function of concentration or duration of exposure.

- Carcinogen: A material that either causes cancer in humans, or, because it causes cancer in animals, is considered capable of causing cancer in humans.

- Toxic chemicals are chemicals that are poisonous to living organisms when they are ingested, inhaled, or absorbed through the skin. In sufficient quantity, any chemical—including such common ones as water and table salt—can be toxic. However, the factor that determines whether a product is considered acutely toxic is the dose (the amount ingested, inhaled, or absorbed) that is required to cause harm. For chemicals not considered toxic, the dose required to produce toxic effects is so high that it is unlikely to be met except under extraordinary circumstances. Substances that are considered acutely toxic can produce harmful effects after a single dose. Examples of acute toxicity include burns and poisoning. The smaller the dose required to produce a toxic effect, the more toxic a chemical is considered to be. Thus, "the dose makes the poison."

Chemicals that do not pose an immediate risk but rather a long-term risk are called chronically toxic, or are said to cause chronic toxicity. These chemicals are generally not harmful in single doses, but they cause harm when organisms are exposed to them over long periods of time. Individual doses are usually too small to cause immediate acute effects; instead harmful effects appear after a period of time. Examples of chronic effects include liver or kidney damage, central nervous system (CNS) damage, cancer, or birth defects.

Toxic chemicals can enter the human body in three major ways: inhalation, ingestion, and dermal exposure. Inhalation can involve breathing contaminated air or contaminated particles of soil and dust. Ingestion includes swallowing contaminated materials. Dermal exposure is contact with the skin.

By law, hazardous products must bear labels that explain the hazards associated with them and how to prevent injury or damage. Signal words determined by law express the relative risk associated with a product. These signal words are "warning" or "caution," "danger," and "poison." The term "nontoxic" is an advertising word that has no legal meaning except when used to describe art supplies. The following is an explanation of the signal words:

- No signal word—relatively nonhazardous

- Caution or Warning—generally mildly to moderately hazardous or toxic; can cause temporary adverse health effects, such as skin irritation or vomiting

- Danger—more severely hazardous or toxic; can cause permanent serious health effects, such as skin burns or stomach ulcers

- Poison—highly toxic; can be fatal if ingested

The Student Information sheet contains a list of various chemicals, signal words, some products in which they are likely to be found, and the doses required to produce toxic effects.

Who Produces Hazardous Waste?

About 15,000 industrial generators of hazardous waste in the United States collectively produce more than 264 million metric tons of hazardous waste annually. Only an estimated 10% of this hazardous material is being disposed of in an environmentally safe manner. However, industries are not the only disposers of hazardous wastes. Hazardous products can be found in almost every household in the United States. When these products are improperly used or disposed of, their hazardous chemicals can be released into the environment.

According to *Garbage* magazine, each family in the United States produces about 7 kg of hazardous waste per year—that's 7 million kg per year just from households. Hazardous materials can be found in many areas of the home. In the garage, hazardous chemicals are found in antifreeze, brake fluid, wax polish, engine degreaser, carburetor cleaner, creosote, radiator flushes, asphalt, roofing tar, air conditioner refrigerants, and car batteries. A work bench can house rust preventatives, wood preservatives, wood strippers, wood stains, paint thinner, oil-based paint, solvents, degreasers, sealants, and varnish. In the house, hazardous chemicals are found in drain cleaners, oven cleaners, furniture polish, metal polish, window cleaners, expired prescriptions, arts and crafts supplies, photography chemicals, floor cleaners, chemistry sets, mothballs, and rug and upholstery cleaners. The garden shed can contain pesticides (including chlordane), herbicides, insect sprays, rodent killers, pool chemicals, insect strips, fertilizers, and septic system cleaners.

Table 1 shows the percentages of household products that may be considered hazardous when they enter the waste stream.

Table 1: Percentage of Household Hazardous Waste by General Type	
household maintenance items (mostly paint-related)	36.6%
household batteries	18.6%
personal care products	12.1%
cleaners	11.5%
automotive-maintenance products	10.5%
pesticides, pet supplies, and fertilizers	4.1%
hobby/other	3.4%
pharmaceuticals	3.2%

What Problems Are Associated with Hazardous Waste Disposal?

Hazardous wastes should never be disposed of as regular trash. Hazardous waste can injure trash collectors as they are picking it up, transporting it, or dropping it off at the landfill. One study showed that 3% of California's trash collectors had been hurt by handling hazardous items that should not have been mixed in with regular trash in the first place. Also, because many hazardous wastes can dissolve plastic and steel containers, improperly stored or discarded waste can pollute the surrounding air, water, and soil.

Disposal of hazardous materials can pose problems for human health and the environment. Commonly, household hazardous waste is mixed in with regular waste and sent to a sanitary landfill or an incinerator. Sometimes people dispose of hazardous wastes themselves by burning them, dumping them onto the ground or into waterways, or by pouring them down a sink, toilet, or storm sewer. All of these methods can potentially be harmful to human health and the environment. In most cities and towns, hazardous chemicals poured down the sink or flushed down the toilet go to a sewage treatment plant. Some of these chemicals are removed or treated to make them less harmful, but others may not be, because most municipal sewage treatment plants are not set up to remove a broad spectrum of hazardous contaminants. These chemicals can then be discharged into waterways. Hazardous wastes flushed into a septic tank spread through the drainfield and can potentially enter groundwater or waterways. Hazardous waste dumped on the ground can be carried by precipitation into storm drains, waterways, or groundwater. Even a small amount of certain chemicals can contaminate large amounts of water. One liter of motor oil disposed of improperly can pollute nearly 950,000 L of water.

Hazardous chemicals should be disposed of in a way that will prevent them from coming into contact with humans or the environment. Ordinances regulating disposal vary from place to place, so before disposing of hazardous products, people should find out what their local ordinances require. Possible sources of information include city, county, or state government; local solid waste districts; soil and water conservation districts; or local waste disposal companies. Some communities hold special collection days for household hazardous waste, and some businesses accept certain kinds of products for recycling. The following are options that may be available for recycling or reusing certain products:

- paint—may be donated to local theater groups, charities, or churches, or may be accepted by hardware stores (Lead-based paint should never be donated; instead, it should be disposed of as hazardous waste.)

- watch batteries—may be accepted by jewelry stores

- hearing-aid batteries—may be accepted by places that sell hearing aids

- nickel-cadmium batteries—may be accepted by cellular phone companies, or call 1-800-BATTERY

- used motor oil—may be accepted by service stations

- cleaning supplies—may be donated to local charities or churches

- fluorescent light bulbs—may be accepted by a company that recycles lamps and ballasts

Toxic and hazardous waste disposal must follow specific guidelines according to federal laws such as Comprehensive Environmental Response, Compensation, and Liability Act (CERCLA); Resource Conservation and Recovery Act (RCRA); and Toxic Substances Control Act (TSCA).

		Household Hazardous Waste		
Products	Before Choosing or Using	To Safely Use Some Products May Require...	If You Can't Use or Share Leftovers...	Potential Hazard Properties
Home				
Adhesives/Glues/ Epoxy	Use water-based when possible.	Fan, safety goggles, gloves	Save liquids for your community's household hazardous waste (HHW) pick-up. Dried products can be put in the trash; call your local environmental agency first to verify this is acceptable.	Flammable, Irritant, Toxic
Aerosols	If you choose aerosols, use entire can or share leftovers. Aerosols no longer contain CFCs.	Don't breathe vapors. Fan, no flames.	Place empty containers in trash, not a trash compactor, or recycle if possible. If can contains product, call your local environmental agency for options.	Flammable
Asphalt/Roofing Tar/Driveway Sealers	Do not allow product to run into storm drains; consider having driveway sealed by professionals.	Use caution to avoid burns. Fans, gloves, safety goggles, protective clothing.	Save liquids for your community's household hazardous waste (HHW) pick-up. Call your local environmental agency about options for dried products.	Combustible, Toxic
Household Batteries	Consider rechargeable batteries. They can be recharged and recycled.	Keep button batteries away from children.	For "Ni-Cd" rechargable batteries, call 1-800-BATTERY. For all others, call your local environmental agency for options.	Corrosive, Toxic
Fluorescent Lights	Fluorescents save energy; choose low-mercury if available.	Avoid breaking fluorescent lights; do not open ballasts.	Call your local environmental agency for options.	Toxic
Mercury Thermostats/ Thermometers	Consider electronic/digital thermostats or thermometers.	Call local fire department for advice on cleaning up mercury spills.	Ask manufacturer if they accept used mercury thermometers. Otherwise, save for your community's household hazardous waste (HHW) pick-up.	Toxic
Mothballs	Clean clothes before storing in sleeved container with cedar chips. Do not use mothballs continuously.	Gloves, fan, air out exposed clothes before use.	If container is less than ¼ full, wrap closed, original container in newspaper and place in the trash. Save larger quantities for your community's household hazardous waste (HHW) pick-up; call your local environmental agency first to verify they will accept this product.	Irritant, Toxic
Undiluted Pool Chemicals	Buy only as much as you need. Ask retailer for advice on quantities.	Fan, gloves, safety goggles, protective clothing	Save for your community's household hazardous waste (HHW) pick-up; call your local environmental agency first to verify they will accept this product.	Corrosive, Irritant, Toxic
Propane Gas Cylinders	Ask retailer about a propane cylinder exchange program; Make sure cylinder connection matches appliance.	Use 20-lb. gas grill cylinders outdoors only.	Return 20-lb. propane cylinders (gas grill size) to a local propane retailer who accepts them. Call your local environmental agency for options for smaller cylinders.	Flammable

	Household Hazardous Waste, continued			
Products	Before Choosing or Using	To Safely Use Some Products May Require...	If You Can't Use or Share Leftovers...	Potential Hazard Properties
Cleaners/ Polishes				
Bleach	Color-safe bleach is an alternative to chlorine bleach but has less whitening strength and is not a registered disinfectant.	Fan, gloves, safety goggles. Do not mix with other cleaning materials.	Pour small amounts of liquid (less than one cup) into the toilet, after you remove any bowl cleaners and deodorizers, and flush. Allow at least one day between disposal of different products. Never pour in storm drain. For powder, put sealed box in a bag and place in the trash.	Irritant
Drain Openers	Use a drain filter; do not pour grease down drain. Use a snake/plunger to unclog.	Fan, gloves, safety goggles. Do not mix with other cleaning materials.	Pour small amounts of liquid (less than one cup) into the toilet, after you remove any bowl cleaners and deodorizers, and flush. Allow at least one day between disposal of different products. Never pour in storm drain. For crystals, if container is less than ¼ full, wrap closed, original container in newspaper and place in the trash. Save larger quantities for your community's household hazardous waste (HHW) pick-up; call your local environmental agency first to verify they will accept this product.	Corrosive, Irritant
Oven Cleaners	Put baking sheet on lowest oven rack to catch spills; try non-corrosive or fume-free products.	Fan, gloves, safety goggles. Do not mix with other cleaning materials. After use, leave room until fumes subside.	Place empty aerosol cans in trash (not trash compactor) or recycle if available. If can contains product, call your local environmental agency for options. For non-aerosols, if container is less than ¼ full, wrap closed, original container in newspaper and place in the trash. Save larger quantities for your community's household hazardous waste (HHW) pick-up; call your local environmental agency first to verify they will accept this product.	Corrosive, Irritant
Tub/Tile/Sink	Wipe down tub and sink after use to prevent buildup.	Fan, gloves, safety goggles. Do not mix with other cleaning materials.	Pour small amounts of liquid (less than one cup) into the toilet, after you remove any bowl cleaners and deodorizers, and flush. Allow at least one day between disposal of different products. Never pour in storm drain. Place empty aerosol cans in trash (not trash compactor) or recycle if available. If can contains product, call your local environmental agency for options.	Corrosive, Irritant

Household Hazardous Waste, continued				
Products	Before Choosing or Using	To Safely Use Some Products May Require...	If You Can't Use or Share Leftovers...	Potential Hazard Properties
Cleaners/ Polishes, cont.				
Disinfectants	Choose a product with the words "EPA REG. NO." on the label.	Fan, gloves, safety goggles. Do not mix with other cleaning materials.	Pour small amounts of liquid (less than 1 cup) into the toilet, after you remove any bowl cleaners and deodorizers, and flush. Allow at least one day between disposal of different products. Never pour in storm drain. Place empty aerosol cans in trash (not trash compactor) or recycle if available. If can contains product, call your local environmental agency for options.	Corrosive, Irritant, Toxic
Toilet Bowl Cleaners	Consider cleaners labeled Noncorrosive.	Fan, gloves, safety goggles. Do not mix with other cleaning materials.	Pour small amounts of liquid (less than 1 cup) into the toilet, after you remove any bowl cleaners and deodorizers, and flush. Allow at least one day between disposal of different products. Never pour in storm drain. For non-liquids, if container is less than ¼ full, wrap closed, original container in newspaper and place in the trash. Save larger quantities for your community's household hazardous waste (HHW) pick-up; call your local environmental agency first to verify they will accept this product.	Corrosive, Irritant
Furniture Polish	Consider products labeled Caution or Warning instead of Danger when purchasing.	Fan, gloves, safety goggles.	Place empty aerosol cans in trash (not trash compactor) or recycle if available. If can contains product, call your local environmental agency for options. For non-aerosols, if container is less than ¼ full, wrap closed, original container in newspaper and place in the trash. Save larger quantities for your community's household hazardous waste (HHW) pick-up; call your local environmental agency first to verify they will accept this product.	Flammable, Irritant, Toxic

When considering whether to buy a hazardous product, consumers should consider the following questions:

- Do I really need this product? Would a less hazardous product or procedure work satisfactorily?

- Am I willing to take responsibility for storing and disposing of a product I need to use?

- Do I really need to buy this much? Can I use it all before it expires?

- Will the store let me try a small amount before I make this purchase?

- Can I use a liquid form of this product, one that brushes on, rolls on, or can be sprayed from a pump instead of an aerosol can?

- Am I sure I like the color, texture, or other important qualities?

This activity does not contain any recipes or instructions for making homemade alternatives to hazardous products. Homemade products are not necessarily good alternatives, for the following reasons:

- Homemade products may be less effective than commercial products, causing the consumer to spend more time and use more of the product.

- Homemade products may not have been tested for safety or environmental impact.

- Homemade products may not be stored in child-resistant or pet-resistant containers. Storing them in containers previously used for commercial products may pose unexpected hazards if the ingredients of the homemade product react with residue from the commercial product.

- Homemade products may not be labeled with all of the information required for commercial products, such as ingredients and safety and disposal procedures. It may not be possible to tell whether the mixture is toxic, corrosive, or dangerous if mixed with another substance.

- Ingredient information may not be registered with poison control centers. Homemade products have no company consumer information phone number listed on the label to help consumers provide fast treatment in case of an accident.

● Teacher Notes for "Household Hazardous Waste"

In this activity, your students will learn how to assess the hazards of various products by reading product labels, and they will conduct an inventory of hazardous products in their own homes.

Group Size .... 4–5 students
Time Required .... 10–15 minutes for Part A
20 minutes for Part B plus overnight for home inventory

Materials

For the Procedure
Per class
* wide variety of hazardous household products in original packages with clearly legible labels

 You may wish to use empty packages, especially for very hazardous items.

Getting Ready

Rinse empty containers thoroughly, and replace caps or lids if appropriate. For containers that are not empty, make sure their caps or lids are on tightly. Make sure the product labels are completely legible.

Safety and Disposal

Instruct students not to open the containers of the products while reading labels.

Opening Strategy

For Part A: Learning the Label Lingo

1. Ask students what the word "hazardous" means. *Dangerous.*

2. Focus the discussion on "dangerous chemicals." Ask students what kinds of chemicals they typically consider to be "dangerous." Under what conditions are they dangerous? After students have offered some suggestions, point out that hazardous chemicals can be classified by various characteristics: explosive or reactive, flammable/combustible, corrosive, irritant, and toxic or poisonous. Students may tend to focus on only one or two of these characteristics; if necessary, guide the discussion to include all five.

3. Ask students what the words "toxic" and "poisonous" mean. *A toxic or poisonous chemical is capable of causing illness, injury, or death if ingested, inhaled, or absorbed through the skin.* Discuss the idea that all chemicals—even common ones like water or table salt—could potentially be toxic if consumed in sufficient quantity.

4. Point out that we do not refer to table salt as a toxic chemical, even though it has the potential to cause harm. Ask students what quality might determine whether a chemical is usually considered toxic or not. *A chemical is considered toxic if a relatively small amount can cause illness, injury, or death.* Introduce the term "dose" and the expression "the dose makes the poison."

5. Ask students where hazardous materials are likely to be found. Do they think that hazardous materials are found only in industrial settings? Help them to understand that hazardous materials are found in almost every home in the U.S. Have students brainstorm a list of household products they think might be hazardous. (See the Household Hazardous Waste table in the Teacher Background section for this activity for a list of common materials.)

6. Ask students who is responsible for disposing of household hazardous waste. Help them to understand that people are responsible for the household hazardous waste in their own homes and that when they purchase a hazardous product, they assume responsibility for its handling and disposal.

7. Discuss the idea that product labels can provide information about whether a material is hazardous and, if so, what kind of hazard it poses. On the chalkboard or a large chart, list warning words that are likely to be found on hazardous products. These words include "explosive" or "reactive"; "ignitable," "flammable," or "combustible"; "corrosive"; "irritant"; and "toxic" or "poisonous." Approved signal words that are used to indicate how hazardous a product is include "poison," "danger," and "warning" or "caution." These signal words are required by law to appear on labels of hazardous products, and they indicate different levels of hazard. Some labels may say "nontoxic," but this word is an advertising word that has no legal meaning except when used to describe art supplies. Labels for some products, such as batteries, may not contain such words, but they may contain special instructions for disposing of the product. Special disposal instructions are a clue that the product may contain hazardous materials.

8. Distribute the Student Instructions and the Student Information (provided). Tell the students that the Home Pollution Prevention Plan section is to be filled out at home. After students have read the labels of their group's products and recorded their observations, have them rank all of the products according to the signal words. First, ask all of the groups who have "1's" in their charts to come

forward and write their products on a class chart. Have groups with "2's" list those products next, then the "3's" and "4's." If any of the groups had products with no signal words on the label but special disposal instructions, have them list those products as well, along with the special instructions.

For Part B: Developing a Home Pollution Prevention Plan

1. Brainstorm a list of areas found in most homes (such as kitchen, bathroom, garage, basement, lawn). Tell students that they will conduct an inventory of hazardous materials likely to be found in their assigned areas. They will use the information they collect to develop a Home Pollution Prevention Plan as a class.

2. Divide the students into teams of three or four and assign each group an area to inventory. If necessary, shift students to different groups according to the kinds of places they live. For example, students who live in apartments may not be able to inventory garages, basements, or lawns.

3. Instruct all of the students to take home the Home Pollution Prevention Plan section of the Student Instructions, fill it out, and bring it back to class.

Discussion

1. After students have conducted their inventories, pool results in a class chart organized by area. Discuss whether any products overlap areas. Does one area seem to contain an exceptional variety of hazardous materials? What special storage or disposal problems might this area present? Consider such factors as the temperature range of the area, whether products are accessible to children or pets, and whether the products are likely to be completely used up. Compare the class chart to the Household Hazardous Waste table and look for any important items that students may have overlooked. Add these items to the list.

2. Have the class conduct research on waste disposal and recycling options for the household products on the class list. You may wish to invite a representative of your local solid waste district, department of environmental services, or local solid waste company to your classroom for the students to interview and find answers to the questions below. Explain your project carefully so your visitor can be prepared. Alternatively, you could assign as homework the task of calling local businesses and governmental organizations to gather information about current waste disposal methods, recycling options, and local disposal ordinances.
   - Where does your household waste go? (for example, landfill, incinerator)
   - Where does your household's wastewater go? (for example, municipal water treatment plant, septic system) Does this destination affect options for disposing of hazardous liquids?

- What disposal ordinances and recycling opportunities are available in your community? These will vary from place to place, so students will need to contact various organizations to find the information they need. They may start by calling different agencies, such as city government, state government, the local soil and water conservation district, or local environmental organizations.
- Do local community, charitable, or religious organizations accept donations of reusable products such as paint or cleaning products? Are any restrictions placed on donations?
- Do local businesses accept hazardous waste, such as watch batteries, car batteries, used motor oil, or paint?
- Does your community have a special day when household hazardous waste is collected to be placed in a secure landfill?
- Should any products receive special treatment before they are thrown in regular trash?

3. Have students use the information they have gathered to develop a class Home Pollution Prevention Plan for each area of the house. Have them decide as a class which disposal method they think would be most appropriate, based on the effectiveness of the product, the comparative costs of the product, and/or local disposal ordinances. Would their recommendations be different if a given household contained small children or pets?

Extension

Have students present the results of their Home Pollution Prevention Plan to school administrators or your school's parent-teacher organization.

Resources

Building Environmental Education Solutions, Inc. (BEES), http://www.beesinc.org

Environmental Hazards Management Institute (EHMI). Tools for Community Outreach. Household Hazardous Waste Wheel® and Household Product Management Wheel®. http://www.ehmi.org/home.htm

"Pollution and Solid Waste"; *Investigating Solid Waste Issues;* Ohio Department of Natural Resources: Columbus, OH, 1996, pp B-77–B-94.

● Student Instructions for "Household Hazardous Waste"

In this activity, you will learn about the "signal words" that describe hazardous products. Then you will conduct an inventory of your home to see what hazardous products it contains.

Procedure

Part A: Learning the Label Lingo

Do not open the containers.

Read the labels of the products your teacher has collected. Using the Signal Words and Product Hazards table in the Data Recording section, record the products, the signal words used on their labels, and the product hazards. Use the signal words to rank the products according to the level of hazard they pose, as follows: "Poison"—1; "Danger"—2; "Warning" or "Caution"—3; "Nontoxic"—4. Do labels with no signal words contain special instructions for storage, use, and disposal? If so, put an asterisk (*) in the "Hazard Ranking" column.

Part B: Developing a Home Pollution Prevention Plan

Use the Home Pollution Prevention Plan section of this handout to conduct an inventory of your group's assigned area as follows:

1. List the brand name of each product and the kind of product it is.

2. List the signal word used on each label. For convenience, use the initial of the signal word—P, D, W, C, or N.

3. Describe how the product should be disposed of, according to the label.

Data Recording

Table: Signal Words and Product Hazards

Name of Product	Signal Word	Hazard Ranking (1–4 or *)	Product Hazards (Check all that apply.)				
			Flammable/ Combustible	Toxic	Irritant	Corrosive	Explosive or Reactive

Analysis Questions

a. Might any products listed in the table be difficult for the average household to use up? If so, which one(s)?

_____

_____

_____

b. Were you surprised by the information on any of these labels? If so, how?

_____

_____

_____

c. Household hazardous materials require special care in three areas: storage, use, and disposal. What risks are associated with improper handling in each of these areas?

_____

_____

_____

d. What might constitute "improper" storage? (Consult the labels of the products for hints.)

_____

_____

_____

e. What might constitute "improper" use? (Consult the labels of the products for hints.)

_____

_____

f. What might constitute "improper" disposal? (Consult the labels of the products for hints.)

_____

_____

Home Pollution Prevention Plan

Use the following abbreviations for the categories of product hazards most likely to be found on the labels of household products:

C = Corrosive
E = Explosive
F = Flammable
I = Irritant
T = Toxic

Remember, the signal words are "Poison," "Danger," "Warning" or "Caution," and "Nontoxic."

Table: Home Pollution Prevention Plan

Brand Name and Kind of Product	Signal Word	Category of Hazard	How the Product Should Be Disposed Of

● Student Information for "Household Hazardous Waste"

Table: Warning Levels and Toxicities of Common Household Products			
Warning Level	Toxicity	Lethal Dose for 150-Pound Human	Household Products
No Warning	Practically Nontoxic	more than 1 quart	foods, candies, graphite pencils, eye makeup
	Slightly Toxic	1 pint to 1 quart	dry cell batteries, glass cleaner, deodorants and anti-perspirants, hand soap
Caution	Moderately Toxic	1 tablespoon to 1 pint	antifreeze, automotive cleaners, many detergents, dry cleaners, most fuels, lubricating oils, most stain and spot removers, many disinfectants, floor polish, shoe polish, most paints, most oven cleaners, many general cleaners
Warning	Very Toxic	1 teaspoon to 1 tablespoon	toilet bowl cleaners, some deodorizers, engine motor cleaners, some fertilizers, some paint brush cleaners, some paint and varnish removers, fireworks, some mildew proofing, air sanitizers, some paints, lacquer thinners, many pesticides (chlordane, heptachlor, lindane, mirex, diazinon, etc.)
Danger	Extremely Toxic	1 drop to 1 teaspoon	some insecticides, fungicides, rodenticides, and herbicides
	Super Toxic	1 drop or less	a few pesticides such as paraoxon and phosdrin

Understanding Garbage and Our Environment

The Garbage Gazette

May 1   Local Edition   Vol. 1, Issue 12

A Timeline of Trash Disposal

What would your neighborhood be like if garbage collectors didn't exist?

Billions of people from the past and present know about this scenario first-hand. Public waste collection is relatively new to the waste management field, and many cities today have no public waste collection system. Yet, humans have been generating garbage and trash since they appeared on Earth. What have they done with it?

The earliest people followed the herds of the migrating animals they hunted for food and had no permanent settlements. They left behind their garbage when they moved on. The animal and plant matter decomposed and began to smell, but by then, the people had moved away.

Climate changes and the extinction of large game animals forced people to rely more on plants. They learned to cultivate crops, hunting only local animals. Their settlements became permanent, so they had to find new ways to dispose of their garbage.

At first, many people just threw their wastes (including human excrement and animal manure) into the dirt roadways. If small scraps of food and other wastes fell on dirt floors inside the house, the scraps were trampled into the floor. When the roads and floors started to smell of rotting garbage, people spread a new layer of soil over the old to eliminate the odors.

Join the "Loose Paper Haters." Trash cans from the early 1900s with slogans to help encourage litter-free streets.

New layers of soil were added as needed, raising the street and floor levels a little bit each time. Those "little bits" added up: The ancient city of Troy rose 4.7 feet each century, and the street level in Manhattan today is 14 feet higher than it was in the early 1600s.

Some early civilizations used other methods. For example, the Mayans dumped their garbage in large piles on the outskirts of their cities. These "open dumps" were located downwind so the smell of rotting garbage was carried away from the cities.

Ancient Greeks also piled their garbage outside of town, but they covered their piles every day with fresh soil, thus creating the first "landfills." On the nearby island of Crete, residents of Knossos made compost piles: mounds of leftover animal and plant matter that were stirred every few days. After the matter decomposed, it was used as fertilizer.

The first "garbage collectors" were organized in ancient Rome. Signs hung throughout the city asked people to take their trash to the open dumps outside of town instead of throwing it in the streets. The garbage collectors picked up any trash remaining in the streets and swept paved roadways.

Gradually, the practice of using open dumps spread. A tract of unused land outside of town was set aside for the piles of discards from the city's residents. People were supposed to cart their own garbage to the dumps and leave it there, but most people continued to throw their garbage into the streets.

In the 14th century, European towns began hiring "scavengers" who collected garbage from the streets and hauled it to the dumps. They were "paid" with the privilege of having first choice of any objects of value they found among the discards.

People who lived in rural areas disposed of their garbage close to their homes for convenience and safety. Open dumps were undesirable because of the odors associated with them, so trash pits were dug in the yard instead. Garbage was thrown into the pit and covered with a thin layer of dirt to help contain odors. Filled pits were covered with dirt, and new pits were dug as necessary.

The Industrial Revolution of the 18th and 19th centuries caused many changes, including a sudden population growth in the cities. People flocked to the cities to find jobs in the new factories. As cities grew bigger, multistory homes and buildings were built to provide more room for the new residents. People living and working on the upper floors found it easiest to throw their garbage—including the contents of chamber pots, early versions of toilets—out the window and into the streets, most of which were still unpaved. Benjamin Franklin created the first street cleaning system for paved roads in 1757, but it wasn't until the 1800s that cities began paving every street to make cleaning easier.

Also during the 1700s and 1800s, people realized the correlation between garbage and disease. Waste management became a public health issue, and gradually city governments organized more advanced garbage collection systems, including water treatment and sewage systems. The solid waste collected by these systems was usually taken to an open dump. In the late 1800s and early 1900s, people developed alternatives to open dumps, which were too close to the expanding cities.

The first alternative was the incinerator, created in 1874. Incinerators burned garbage and turned it into ash, which was then buried. The smell of burning garbage was very unpleasant, though, and the popularity of incinerators fluctuated as other options were explored. During the energy crisis of the 1970s, incinerators returned as "waste-to-energy" plants that harvested the energy from burning garbage without the odors, thanks to technological advances that provided new air filtration and purification systems. Many cities today are powered by electricity generated by incinerators.

A second alternative was the reduction plant, which gained popularity in the 1880s. Wet garbage—including dead animals from the cities (there were 15,000 dead horses in New York City alone each year)—was stewed in pots to produce grease and a substance called residuum. The grease was sold to make soap, candles, and perfume, and the residuum was sold as fertilizer. The smell from reduction plants was no better than the smell from the early incinerators, and most reduction plants were closed during the 1920s. The last reduction plant in America closed in 1959.

The third alternative to open dumps is the sanitary landfill, which is the most popular option today. Sanitary landfills were used in America as early as 1904 and were made popular in Britain in the 1920s. Sanitary landfills are different from open dumps in two important ways. First, the bottom of a sanitary landfill is lined to prevent liquids from leaching into the soil and nearby waterways. Second, sanitary landfills are covered with soil each day. This keeps the amount of biodegradation in the landfill to a minimum and helps eliminate odors.

Despite what humans have learned about waste management, garbage piles up along city streets, and waterways are contaminated, especially in developing nations. In America, many people are concerned about the amount of garbage we produce and whether we will run out of landfill space. As the world population expands, more people are around to produce garbage. What will we do with it? What does the future hold for our trash? Only time will tell.

Think About It

1. Describe the three alternatives to open dumps for disposing of garbage.

2. Where does your trash go? To a landfill, to an incinerator, or to some other facility? Is your electricity generated by a garbage incinerator? How can you find out?

3. Can you think of other methods for disposing of garbage besides open dumps and the alternatives from question 1? Describe your ideas.

Lesson 8:
Conclusion

Students conclude their exploration of solid waste with two activities. The first, "**Life Cycle Assessment**," brings solid waste issues into a larger context by looking at energy use and environmental impacts throughout the life cycle of a product or material. The second activity, "**Trash Trouble in Tacktown, Part B**," finishes the role-playing activity started in Lesson 1. Students will be able to apply concepts and information from throughout the book to this closing experience.

Students will have an opportunity to use what they have learned as they look at commonly held myths about solid waste in *The Garbage Gazette* "**Garbage Fact and Fiction**."

● Teacher Background for "Life Cycle Assessment"

"Paper or plastic?" is a familiar question in the grocery store check-out line, and many consumers want to make an environmentally sound choice. But on what basis do we decide? Solid waste disposal issues are part of the story, but what about resource use and pollution associated with product manufacturing? An evaluation technique called life cycle assessment provides a complete picture of the energy used and waste produced over the whole life cycle of a product—every step from the acquisition of raw materials to final disposal. This type of evaluation is often referred to as the "cradle-to-grave" approach to resource management. Companies use life cycle data to make improvements in their manufacturing processes. This information can also help consumers select products that use less energy during manufacturing, produce less air or water pollution, or create less solid waste. Life cycle assessment can also show whether or not increased recycling rates will result in decreased energy use or pollution.

Life cycle assessment is a developing discipline. In an effort to provide guidelines for the process, the Society of Environmental Toxicology and Chemistry (SETAC) and the United States Environmental Protection Agency (EPA) have defined three components of a complete life cycle assessment: life cycle inventory, life cycle impact assessment, and life cycle improvement assessment. The life cycle inventory identifies and quantifies material usage, energy requirements, solid wastes, and atmospheric and waterborne emissions throughout the life cycle of a product or material. All steps are evaluated: the extraction or acquisition of raw materials; processing into intermediate materials; fabrication into the finished product; and final disposition, including recycling, reuse, composting, or other waste reduction or diversion techniques. Transportation of materials between each step is also included. To organize all of this information, a product's life cycle inventory is often presented as a flow chart. (See Figure 1.) Note that pre-consumer recycling occurs when industrial scrap is channeled back into the preparation of materials. Post-consumer recycling may feed back into the manufacture of the same product (closed loop) or into the manufacture of a different product (open loop). In the United States, a leading provider of life cycle inventories is Franklin Associates, a consulting firm whose clients include the EPA as well as many companies and industry groups.

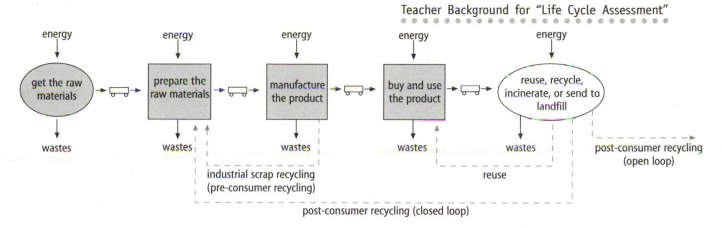

Figure 1: A product life cycle inventory is often presented as a flow chart.

Data from the life cycle inventory are used to complete the next step in the process: life cycle impact assessment (also called inventory interpretation). Many impacts may be evaluated, such as air and water quality, human health, resource consumption, global warming, and biodiversity. Currently, no accepted strategy exists for combining all the different impacts of a product to create a single comprehensive impact measure, or score. A comprehensive score for each product appeals to decision makers because it reduces the decision-making process to a simple numerical comparison. However, Franklin Associates warns that simple scoring approaches shield decision makers from the true complexity of impact assessment.

Without a single measure, or score, comparing the impact of one product to another is very difficult. Life cycle assessments of similar products have led to controversial outcomes because of disagreements about which impact categories are most important and how these levels of importance can be reflected in weighted scores. For example, how do you choose the "greener" product when one produces more water pollution but the other takes up more landfill space? The "greener" choice depends on the assessor's preference and local environmental conditions.

Life cycle improvement assessment uses the results of the inventory and impact assessment to suggest changes in the manufacturing process that will reduce environmental impact.

Despite its limitations, life cycle assessment is emerging as an important tool for product design and consumer decision-making. Advocates acknowledge that the process is not perfect, but they point out that life cycle thinking is affecting the environmental awareness of both designers and consumers and leading to "greener" products in the marketplace.

● Teacher Notes for "Life Cycle Assessment"

In this activity, students will review life cycle inventory data for several common products. When finished, students may have a different perspective on which products have less environmental impact.

Group Size.. 3–4 students
Time Required .... Getting Ready: 5–10 minutes
Procedure: 20–30 minutes
Discussion: 25–35 minutes

Safety and Disposal

No special safety or disposal procedures are required.

Materials

For the Procedure
Per class
- clean product samples
  ○ paper grocery bag
  ○ plastic grocery bag
  ○ Styrofoam™ cup
  ○ wax-coated paper cup
  ○ 1-gallon plastic milk jug
  ○ ½-gallon paper milk carton
- overhead transparency of life cycle inventory flow chart (Figure 1 in Teacher Background)
- overhead transparency of blank life cycle inventory flow chart from the Data Recording section of the Student Instructions

Opening Strategy

1. Put the product samples where they are visible to the class, discuss the following questions with your students, and ask students to explain the reasons for their choices.
   - Which type of bag do you or your families prefer to receive at the supermarket—paper or plastic?
   - If given the choice between a foamed polystyrene cup and a wax-covered paper cup, which would you choose?
   - How does your family usually buy milk—in a 1-gallon plastic jug or a cardboard ½-gallon container?

2. Introduce the concept of life cycle assessment as a tool to help consumers and manufacturers make informed choices about products. Bring out the idea that solid waste issues are one important part of a product's life cycle but that energy use and other environmental impacts are important too.

3. Explain the three stages of a life cycle assessment. Point out some of the benefits and limitations of the process. Tell students that they will be focusing their attention on the first stage of the assessment process—life cycle inventory.

4. Put the life cycle inventory flow chart transparency (Figure 1 from the Teacher Background) on the overhead projector, and cover it completely with a piece of paper. Ask students what they think the first step in the life cycle of any product would be. After some discussion, reveal the first oval in the flow chart (get raw materials) and discuss its meaning. Point out the arrows representing energy inputs and waste outputs.

5. Continue in a similar fashion through the other four steps of a life cycle inventory. Have students suggest ways that energy might be consumed and waste products produced at each stage. Make sure that students are aware that energy is used for transporting raw materials, products, and solid waste and that waste is produced in the form of air emissions, water effluents, and solid waste. Explain that the measures of energy use and waste produced through the entire life cycle of a product is a life cycle inventory.

6. Point out that pre-consumer recycling occurs when industrial scrap is channeled back into the preparation of materials. Explain that post-consumer recycling may feed back into the manufacture of the same product (closed loop) or into the manufacture of a different product (open loop). For example, recycled plastic can be used to make products such as flower pots, carpets, and fabric.

7. Put a transparency of the blank flow chart (from the Data Recording section) on the overhead projector. Through a class discussion, fill in the flow chart for aluminum cans. (See sample in Figure 2.)

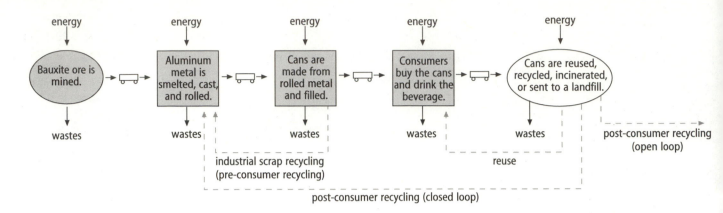

Figure 2: At the end of the class discussion, your life cycle flow chart for aluminum should be similar to this example.

8. Tell students that they will use life cycle inventory data to examine the relative environmental impacts of the products they discussed in step 1, and explain the following: The life cycle inventory data were collected by Franklin Associates (a leading provider of life cycle inventories in the United States). No matter how carefully data are collected and analyzed, some uncertainty always exists about how close the measurements are to the real values. This uncertainty is represented by a number called margin of error. According to Franklin Associates, the margin of error for the energy and solid waste data is about plus or minus 10%. For the air and water impacts, the margin of error is about 25%. This is important, because if the difference between two products' measures for energy use or solid waste produced is less than 10%, the difference may just be due to error in the numbers. If the difference between two products' measures is greater than 10%, we can be confident that the difference is real.

9. Tell students that the data provided to them already includes percent differences, but that you want them to see how simple the calculation is. Show students how to calculate the percent of difference between two numbers by applying the following formula to a data sample from one of the Student Information sheets.

$$\frac{larger\ number - smaller\ number}{larger\ number} \times 100 = \%\ difference$$

10. Divide the class into groups of 3–4 students. Assign each group one of the product pairs to investigate: paper and plastic grocery bags, foam polystyrene and waxed paper cups, or plastic and cardboard milk containers. Give students a copy of the Student Instructions and the Student Information for their product pair, and have them begin the Procedure.

Discussion

1. Have the small groups for each product pair join to form a large group. (You will now have three large groups—one for each product pair.) Instruct the large groups to plan a 5- to 10-minute presentation to the class about the life cycle inventory for their products. Have students select one representative from each of the original small groups to present their information to the class.

2. Allow sufficient time for information about each product pair to be presented to the class.

3. Remind students of the class discussion in step 1 of the Opening Strategy. Ask, "Would you make any different choices based on what you now know?" Discuss the importance of looking at all the evidence and not basing decisions on preconceived notions.

Cross-Curricular Integration

Language arts:
• Have students write an advertisement for one of the products in the activity, using data from the Life Cycle Inventory to support their claims.

Mathematics:
• Have students practice calculating the percent of difference between numbers using the data below.

Combustion Properties of Polystyrene and Paper Hinged Food Containers (per 10,000 uses)			
	Weight (pounds)	Energy derived (million Btu per 10,000 uses)	Ash generated by incineration (pounds)
foam polystyrene	112.3	2.0	1.7
LDPE-coated paper	328.2	2.9	18.5

Social studies:
• Conduct a class discussion about the values our culture associates with using different materials. For example, ask students which they think most people see as "better": using paper products or using plastic ones. Discuss the possible reasons for people's attitudes.

Resources

"Let It Flow"; *Exploring the Environment;* Hamilton County Department of Environmental
Services: Cincinnati, OH, 1996; pp 85–90.

"Retrace Your Waste" and "Paper or Plastic"; *An Ounce of Prevention;* National Science Teachers
Association: Arlington, VA, 1996; pp. 91–95, 97–104.

● Student Instructions for "Life Cycle Assessment"

When you hear "Paper or plastic?" in the grocery store check-out line, how do you decide? A process called life cycle assessment can help you make environmentally sound choices about bags and other products. Life cycle data can also help manufacturers find ways to use less energy and produce less waste while making products. In this activity, you will compare familiar products using life cycle data.

Safety and Disposal

No special safety or disposal procedures are required.

Procedure

1. Read the "Product Overview" on the Student Information sheet for your assigned product pair.

2. Using the information about your product in the Product Overview, fill in both life cycle inventory flow charts in the Data Recording section of this handout (one flow chart for each material). The ovals and squares represent five steps of a product life cycle: getting the raw materials, preparing the raw materials, making the product, buying and using the product, and final destination (including reuse, recycling, combustion, or landfilling.) Add arrows to represent energy input, waste output, reuse, and recycling.

3. Read the Energy Requirements table on your Student Information sheet and answer questions a–b in the Analysis Questions section of this handout. (If your product pair does not have complete recycling data, you will not be able to answer question b.)

4. Read the Air/Water Impacts table on your Student Information sheet and answer questions c–f in the Analysis Questions section. (If your product pair does not have complete recycling data, you will not be able to answer some of the questions.)

5. Read the Combustion Properties table on your Student Information sheet and answer question g in the Analysis Questions section.

6. Read the Landfill Volumes table on your Student Information sheet and answer question h in the Analysis Questions section.

7. Answer questions i–k in the Analysis Questions section.

Data Recording

final destination

buy and use
the product

manufacture
the product

prepare the
raw materials

get the raw
materials

product name:

product name:

Analysis Questions

a. Which of your products uses less energy throughout its life cycle if no post-consumer recycling occurs? Is the difference in energy use between the two products greater than 10%? Why are the percentages of difference important?

_____

_____

_____

_____

b. Does the same product use less energy throughout its life cycle if 50% recycling occurs? What about 100% recycling? Does the percentage of difference change? Does it increase or decrease? Why do you think the amount of energy use changes as the amount of recycling increases?

_____

_____

_____

_____

c. Which of your products produces fewer atmospheric emissions throughout its life cycle if no recycling occurs? Is the difference in atmospheric emissions between the two products greater than 25%?

_____

_____

_____

_____

d. Does the same product produce fewer atmospheric emissions if 50% recycling occurs? What about 100% recycling? Does the percentage of difference change? Does it increase or decrease? Why do you think the amount of atmospheric emissions changes as the amount of recycling increases?

_____

_____

_____

_____

_____

e. Which of your products produces fewer waterborne wastes throughout its life cycle if no recycling occurs? Is the difference in waterborne wastes between the two products greater than 25%?

_____

_____

_____

_____

f. Does the same product produce fewer waterborne wastes if 50% recycling occurs? What about 100% recycling? Does the percentage of difference change? Does it increase or decrease? Why do you think the amount of waterborne wastes changes as the amount of recycling increases?

_____

_____

_____

_____

g. Which of your products produces more energy through combustion? Which produces more ash? Why is the amount of ash important? Are the percentages of difference greater than 10%?

_____

_____

_____

_____

h. Which of your products results in a greater landfill volume? Considering what you know about landfills, which information about discarded material is most important—the weight, the density, or the volume? Explain your answer.

_____

_____

_____

_____

_____

i. Were you surprised by any of the life cycle inventory data for your products? Explain.

_____

_____

_____

_____

_____

j. Consider all of the life cycle inventory data for each of your products. We can consider less energy use, less atmospheric and water waste, less ash produced with incineration, and less landfill volume to be environmental advantages. Does one of your products have an environmental advantage over the other in all areas? If not, explain.

_____

_____

_____

_____

_____

k. Will you change any of your consumer habits based on the life cycle inventory data for your products? Explain what you will change and your reasons for doing so.

_____

_____

_____

_____

_____

_____

_____

● Student Information A for "Life Cycle Assessment"
Paper and Plastic Grocery Bags

The first step in a life cycle assessment is the life cycle inventory. This information sheet presents life cycle inventory data for paper and plastic (polyethylene) grocery bags.

Product Overview

The manufacture of polyethylene grocery bags begins with the extraction of crude oil and natural gas from the earth. These raw materials are transported and then processed to form polyethylene. The polyethylene is transported to factories where it is used to manufacture bags. During the manufacturing process, scrap polyethylene is collected and processed to be reused in the manufacture of more bags. Paper grocery bag manufacturing begins with the harvesting of trees for wood. This raw material is transported and processed to form unbleached kraft paper. The kraft paper is transported to factories where paper grocery bags are made. During the manufacturing process, scrap paper is collected and processed to be reused in the manufacture of more bags. Both types of bags are transported to grocery stores, where they are used by consumers to bring home their groceries. Consumers may reuse, recycle, or throw away either type of bag. Paper bags could be composted. Recycling of materials after use is called post-consumer recycling. During recycling, the bags are processed to prepare their materials for reuse in manufacturing. The materials may be used to make new bags (closed-loop recycling) or completely different products (open-loop recycling). Discarded bags may be incinerated or placed in a landfill.

Data Tables

Each data table presents life cycle inventory data based on 10,000 grocery bag uses. Since it takes about two plastic bags to hold the same amount of groceries as one paper bag, 10,000 uses equals 10,000 paper bags or 20,000 plastic bags. The energy use and environmental impact data are for the entire life cycle of the product, including transportation from one stage to another. Energy use is given using a standard measure called British thermal units (Btu). In some tables, energy use or wastes produced are given for post-consumer recycling rates of 0%, 50%, and 100%. This information is for comparison only and is not meant to say that recycling rates as high as 100% are possible or even desirable.

Energy Requirements for Polyethylene and Paper Grocery Bags (million Btu per 10,000 uses)

	0% recycling	50% recycling	100% recycling
polyethylene	13.0	11.1	9.3
paper	16.3	12.7	9.1
percent difference	20.2%	12.6%	2.2%

Air/Water Impacts for Polyethylene and Paper Grocery Bags (pounds of waste per 10,000 uses)

	0% recycling	50% recycling	100% recycling
Atmospheric Emissions			
polyethylene	23.9	21.1	18.3
paper	64.2	48.2	32.2
percent difference	62.8%	56.2%	43.2%
Waterborne Wastes			
polyethylene	2.4	2.2	1.9
paper	31.2	37.6	43.9
percent difference	92.3%	94.1%	95.7%

Combustion Properties for Polyethylene and Paper Grocery Bags (per 10,000 uses)

	Weight (pounds)	Energy derived (million Btu per 10,000 uses)	Ash generated by incineration (pounds)
polyethylene	329.0	6.6	negligible
paper	1,340.0	9.7	13.5
percent difference	75.4%	32.0%	100%

Landfill Volumes for Polyethylene and Paper Grocery Bags (per 10,000 uses)

	Weight (pounds)	Landfill density (pounds per cubic yard)	Landfill volume (cubic yards)
polyethylene	329.0	670.0	0.5
paper	1,340.0	740.0	1.8
percent difference	75.4%	9.5%	72.2%

● Student Information B for "Life Cycle Assessment"
Foam Polystyrene and Wax-Coated Paper Cups

The first step in a life cycle assessment is the life cycle inventory. This information sheet presents life cycle inventory data for foam polystyrene (often called Styrofoam®) and wax-coated paper cups.

Product Overview

The manufacture of foam polystyrene cups begins with the extraction of crude oil and natural gas from the earth. These raw materials are transported and then processed to form polystyrene. The polystyrene is transported to factories where it is used to manufacture the foam cup product. During the manufacturing process, scrap polystyrene is collected and processed to be reused in the manufacture of more cups. Paper cup manufacturing begins with the harvesting of trees for wood. This raw material is transported and processed to form bleached paperboard. The paperboard is transported to factories where cups are made. During the manufacturing process, scrap paper is collected and processed to be reused in the manufacture of more cups. The wax coating for the paper cups starts as crude oil. The oil is processed to form wax, which is transported to factories where paper cups are made. Both types of cups are transported to restaurants and grocery stores, where they are used by consumers to hold beverages. Consumers may reuse, recycle, or throw away the cups. Recycling of materials after use is called post-consumer recycling. During recycling, the cups are processed to prepare their materials for reuse in manufacturing. The materials may be used to make new cups (closed-loop recycling) or completely different products (open-loop recycling). Discarded cups may be incinerated or placed in a landfill.

Data Tables

Each data table presents life cycle inventory data based on 10,000 cups. The energy use and environmental impact data are for the entire life cycle of the product, including transportation from one stage to another. Energy use is given using a standard measure called British thermal units (Btu). In some tables, energy use or wastes produced are given for post-consumer recycling rates of 0%, 50%, and 100%. This information is for comparison only and is not meant to say that recycling rates as high as 100% are possible or even desirable.

Energy Requirements for Foam Polystyrene and Wax-Coated Paper Cups
(million Btu per 10,000 uses)

	0% recycling	50% recycling	100% recycling
foam polystyrene	5.2	4.5	3.8
wax-coated paper	8.3	not available	not available
percent difference	37.3%		

Air/Water Impacts for Foam Polystyrene and Wax-Coated Paper Cups
(pounds of waste per 10,000 uses)

	0% recycling	50% recycling	100% recycling
Atmospheric Emissions			
foam polystyrene	11.8	11.1	10.5
wax-coated paper	21.8	not available	not available
percent difference	45.9%		
Waterborne Wastes			
foam polystyrene	2.1	1.8	1.4
wax-coated paper	4.5	not available	not available
percent difference	53.3%		

Combustion Properties for Foam Polystyrene and Wax-Coated Paper Cups
(per 10,000 uses)

	Weight (pounds)	Energy derived (million Btu per 10,000 uses)	Ash generated by incineration (pounds)
foam polystyrene	96.9	1.7	1.5
wax-coated paper	287.9	3.2	13.0
percent difference	66.3%	46.9%	88.5%

Landfill Volumes for Foam Polystyrene and Wax-Coated Paper Cups
(per 10,000 uses)

	Weight (pounds)	Landfill density (pounds per cubic yard)	Landfill volume (cubic yards)
foam polyestyrene	96.9	180.0	0.54
wax-coated paper	287.9	800.0	0.36
percent difference	66.3%	77.5%	33.3%

Student Information C for "Life Cycle Assessment"
Plastic and Paperboard Milk Containers

The first step in a life cycle assessment is the life cycle inventory. This information sheet presents life cycle inventory data for 1-gallon plastic (high-density polyethylene) and ½-gallon paperboard milk containers.

Product Overview

The manufacture of high-density polyethylene (HDPE) milk jugs begins with the extraction of crude oil and natural gas from the earth. These raw materials are transported and then processed to form HDPE. The HDPE is transported to factories where it is used to manufacture the milk jug product. During the manufacturing process, scrap polyethylene is collected and processed to be reused in the manufacture of more jugs. Paperboard carton manufacturing begins with the harvesting of trees for wood. This raw material is transported and processed to form bleached paperboard. The paperboard is transported to factories where cartons are made. During the manufacturing process, scrap paper is collected and processed to be reused in the manufacture of more cartons. The low-density polyethylene (LDPE) coating for the paper cartons starts as crude oil and natural gas. These raw materials are processed to form LDPE, which is transported to factories where paperboard milk cartons are made. Both types of containers are transported to dairies, filled with milk, and transported to grocery stores, where they are purchased by consumers. Consumers may reuse, recycle, or throw away the containers. Recycling of materials after use is called post-consumer recycling. During recycling, the containers are processed to prepare their materials for reuse in manufacturing. The materials may be used to make new milk containers (closed-loop recycling) or completely different products (open-loop recycling). Discarded containers may be incinerated or placed in a landfill.

Data Tables

Each data table presents life cycle inventory data based on packaging for 1,000 gallons of milk. Since it takes two ½-gallon paperboard cartons to hold the same amount of milk as a 1-gallon plastic jug, packaging for 1,000 gallons equals 1,000 plastic jugs or 2,000 paperboard cartons. The energy use and environmental impact data are for the entire life cycle of the product, including transportation from one stage to another. Energy use is given using a standard measure called British thermal units (Btu). In some tables, energy use or wastes produced are given for post-consumer recycling rates of 0%, 50%, and 100%. This information is for comparison only and is not meant to say that recycling rates as high as 100% are possible or even desirable.

Energy Requirements for High-Density Polyethylene and Paperboard Milk Containers
(million Btu per 1,000-gallon milk delivery)

	0% recycling	50% recycling	100% recycling
high-density polyethylene	5.5	4.6	3.7
paperboard	7.3	not available	not available
percent difference	24.7%		

Air/Water Impacts for High-Density Polyethylene and Paperboard Milk Containers
(pounds of waste per 1,000-gallon milk delivery)

	0% recycling	50% recycling	100% recycling
Atmospheric Emissions			
high-density polyethylene	10.1	8.7	7.2
paperboard	22.2	not available	not available
percent difference	54.5%		
Waterborne Wastes			
high-density polyethylene	0.9	1.0	1.1
paperboard	3.8	not available	not available
percent difference	76.3%		

Combustion Properties for High-Density Polyethylene and Paperboard Milk Containers
(per 1,000-gallon milk delivery)

	Weight (pounds)	Energy derived (million Btu per 1,000 gallons)	Ash generated by incineration (pounds)
high-density polyethylene	141.5	2.8	0.1
paperboard	284.4	2.5	16.5
percent difference	50.2%	10.7%	99.4%

Landfill Volumes for High-Density Polyethylene and Paperboard Milk Containers
(per 1,000-gallon milk delivery)

	Weight (pounds)	Landfill density (pounds per cubic yard)	Landfill volume (cubic yards)
high-density polyethylene	141.5	355.0	0.40
paperboard	284.4	740.0	0.38
percent difference	50.2%	52.0%	5.0%

● Teacher Notes for "Trash Trouble in Tacktown, Part B"

This activity completes the role-playing scenario introduced in Lesson 1. In this part of the scenario, students play their assigned roles in the simulated Town Meeting.

> *Due to the nature of this activity, no Student Instructions are provided.*

Materials

For Getting Ready
Per class
- heavy paper or cardstock
- scissors

For the Procedure
Per student
- tape or straight pin to attach name tag

Per student with a Presentation Role
- large piece of cardstock or tagboard
- markers

Getting Ready

1. Copy the Name Tags (provided) onto heavy paper or cardstock and cut them apart along the lines.

2. Set up a row of desks or tables and chairs for the mayor, the four town council members, and the town treasurer at the front of the room. Be sure that the town treasurer has access to either a chalkboard or a dry-erase board on which to write the relative economic costs of the different plans.

Procedure

1. Distribute Name Tags to all of the students.

2. As the mayor, open the town meeting by explaining the reason for the meeting (to come up with a solid waste management program) and inviting each of the five presenters to speak for 5 minutes each. After all five have presented their arguments, give individual citizens or groups a turn to speak. Allow each citizen or group 5 minutes. One representative can speak for the entire group, or

different people can present parts of the argument. It is the mayor's responsibility to limit each group to a total of 5 minutes.

3. Impress upon the students that while each group is speaking, they should be listening, taking notes, and writing down questions they would like to ask at the end of the presentations. All questions must be held until the last group is finished speaking. At this time, students may ask questions to clarify other groups' positions. Some groups may choose to make compromises during this time and form new coalitions.

4. After all of the groups have spoken and their questions have been answered, clarify the number and nature of the positions to be considered. Meet with the town council to reach a decision. While making the decision, the council members should be in an area removed from the rest of the class. They may not all agree on the same solution to the problem. In this case, they should first try to find a compromise between the different positions. If they are unable to compromise, they should vote. It is possible that if more than two positions are being considered, none will receive the necessary majority of votes. If that is the case, they should narrow it down to the two most popular decisions and vote on those. The town council must have a majority vote on one plan.
    Remind them to stay in character and try to make the choices their characters would be most likely to make.

5. Meanwhile, if another adult is available, have him or her lead the rest of the class in a discussion of the questions on the Student Analysis Questions handout. (Students should step out of their roles for this discussion.) If another adult is not available to lead discussion during the town council decision-making process, have students write down their responses on the handout and discuss them after the town council members have rejoined the class.

6. When the mayor and town council have reached their decision, have them come back and present it to the class. After the mayor has read a description of the plan that has been chosen, have each person on the council say whether or not he or she voted for the plan. At this point, the council members should also explain briefly the interests and personal concerns of their characters and how these concerns influenced their votes.

Extension

Have students list the scientific information that was included in the group presentations. Which seemed more important, scientific facts or personal interests and concerns? Was the final decision consistent with available scientific knowledge regarding waste management?

Cross-Curricular Integration

Language arts:

- Have each student write an objective article describing the simulation activity and a subjective article for use in the *Tacktown Times.* The objective article should describe the major positions that were expressed. The subjective article should defend the position that the decision was good for the community or defend the position that the decision was bad for the community, based on the point of view of the character each student played. Make sure the subjective article describes what the community might gain or lose as a result of the final decision.

Mathematics:

- Have the class discuss some of the following questions: How did the groups use mathematics (statistics, percentages, costs, or other data) in making their presentations? How could the use of mathematics and the treatment of data strengthen the positions of the citizens? How could the use of clear, visual displays enhance their presentations?

Social studies:

- Have students attend a meeting of your community's decision-making board and note elements of effective presentations. Have them discuss the handling of conflicting viewpoints and comment on the difference between healthy, constructive conflict and destructive conflict.

Explanation

Integrated waste management balances different waste management options and does not rely exclusively on one disposal method. It fits the available waste management options to a city or region's needs. Many factors influence which disposal methods are best for a particular community. Different regions of the country have different environmental concerns, such as available landfill space, transportation costs, energy needs, and natural resources. All factors that affect communities, including economic and political issues, need to be taken into consideration. The important concept students need to grasp is that there is no right or wrong answer, just groups of people making compromises and coming up with solutions to their problems.

B. Ballott Mayor	**I. Hedge** Town Council Member
I. Might Town Council Member	**A. Abacus** Town Treasurer
M. Bee Town Council Member	**D. Digger** Landfill Manager
P.R. Haps Town Council Member	**A. Genn** Recycling Company Representative

N. Mix Material Recovery Facility Representative	**G. Grows** Local Farmer
M. Mulcher Compost Company Representative	**P. Proffitt** New Business Interest
B. Burns Incineration Company Representative	**L. Lite** Power Company Representative
P. Pulp Paper Company Representative	**M. Whett** Water Company Representative
B. Bond Chemical Company Representative	**S. Small** Chamber of Commerce Small Business Representative

G. Green
Environmental Group Representative

D. Gooder
Citizen Supportive of Recycling

N. Nozie
Landfill Neighbor

I. Dell
Unmotivated Citizen

N. Nimby
Neighbor to Proposed Landfill Site

A. Roma
Compost Company Neighbor

C. Choke
Citizen Concerned about Air Quality

N. M. Ploid
Citizen out of Work

I. Gulp
Citizen Concerned about Water Quality

I. Haul
Truck Driver

I. Bury **Representative** **from Hazardous Hole**	**G.O. Vern** **EPA Representative**
L. Law **Police Officer**	**X. Periment** **Scientist**
I. Cure **Hospital Administrator**	

● Student Analysis Questions for "Trash Trouble in Tacktown, Part B"

a. What do you think is the best decision for Tacktown? Do you think one plan is better than all the rest, or would a combination of plans be most beneficial?

_____

_____

_____

_____

b. Which kinds of arguments do you feel carry the most weight: environmental, economic, or personal? How would you weigh these arguments against each other?

_____

_____

_____

_____

c. Do you feel that this scenario was a realistic representation of the decision-making process? What other issues might be important in decisions like this that may not have been covered? Should other environmental factors, economic considerations, or personal viewpoints have been expressed too?

_____

_____

_____

_____

Understanding Garbage and Our Environment

The Garbage Gazette

June 7   Local Edition   Vol. 1, Issue 13

Garbage Fact and Fiction

Myths are great as bedtime or campfire stories, but in some cases they can be damaging. For example, the many myths about garbage, recycling, and landfills have caused the solid waste industry to be misunderstood, which can lead to uninformed, unwise decision-making.

Before you can make informed decisions about solid waste, you need to know the truth behind some of the most common garbage myths.

Myth: Plastics, polystyrene foam, and disposable diapers make up most of our solid waste.

Fact: Fast-food packaging takes up $\frac{1}{3}$ of 1% of solid waste by volume, polystyrene foam no more than 1%, and disposable diapers about 1.4%. If plastics were not available, manufacturers would have to use more wood, glass, and paper, which take up more volume in landfills than plastics. Since plastic is recyclable, many plastics can be kept out of landfills through recycling.

Myth: Biodegradable plastics are beneficial because they biodegrade in landfills and take up less space.

Fact: Modern sanitary landfills are built to discourage biodegradation, which produces foul odors and contaminates leachate. Since materials biodegrade very slowly in landfills, biodegradable plastics retain most of their bulk in landfills.

Myth: America is running out of places to put landfills.

Fact: For a study by Resources for the Future, economist A. Clark

Fast-food packages often get a "bad wrap" when it comes to waste disposal, but are they really as big a problem as many people think?

Wiseman calculated that at current disposal rates, all of America's garbage for the next 1,000 years would fit into a space 120 feet deep and 44 miles square, or about triple the size of Oklahoma City. Even though a single landfill this big would not be feasible, the important fact is that this space is not all that large compared to the amount of land available in the country.

Myth: Americans are producing rapidly accelerating amounts of garbage on a per-person basis.

Fact: Few data exist on the amount of garbage Americans produced in the past, but the little that exists does not support this claim. Some analyses have even shown that we now produce less waste per person than we did in the past.

Myth: Since newspapers are recyclable, they do not take up much space in landfills.

Fact: Newspapers used to be a good money-maker for church and school groups, but market saturation has turned newspaper recycling into a bust. Nowadays, people usually have to pay to have their newspapers taken away, rather than receiving money for them. So instead of recycling them, people often just throw them away.

Myth: Recycling is a good thing, so people should recycle everything they can.

Fact: The recycling process uses energy and produces waste, some of which is hazardous, just like any other manufacturing process. Recycling, like any other business, is driven by money. Someone has to collect recyclable materials and process them into other things. If this process is too expensive, it drives up the price of the recycled products. It then becomes cheaper to use virgin goods. Also, as with any other business, the recycling market can be saturated, which means recyclers are receiving more goods to recycle each day than they can process. This leads to a surplus, and these surplus materials must be stored, which costs more money than the materials will be worth once they are processed. Market saturation for most recyclable goods fluctuates month to month. Aluminum cans may be in demand by recyclers in August, but by September, the industry may have more cans than it can process. The excess cans must be either stored or disposed of in another way. The newspaper recycling industry has been saturated for years because newspapers are expensive to recycle. Before the paper fibers can be recovered, the paper must first be deinked. The deinking process produces toxic waste, which is expensive to dispose of. In addition, technology has improved how cardboard and other paper products are recycled, which makes recycling them a better—and cheaper—alternative to recycling newspaper.

Myth: Modern packaging practices, including the increased use of plastics, have increased the amount of solid waste in the U.S.

Fact: Even though the use of plastics has increased, plastics have gotten thinner and lighter, which means that less plastic needs to be used per package. In addition, packaging processes have helped solve some solid waste problems. For example, foods packaged in plastic last longer and stay fresher than they would if they were not packaged in plastic, thus reducing the amount of food waste. In addition, food packaging has allowed for more processed foods, like cans of pre-cut vegetables and packages of deboned meats, that reduce the amount of household food waste (stems, stalks, cores, bones, etc.) that is usually thrown away. Food manufacturers collect these food scraps and sell them to farmers as feed for livestock or as compost.

Myth: Fast-food restaurants should use paperboard containers instead of polystyrene because paperboard is more environmentally friendly.

Fact: This debate has an interesting history. In the 1970s, McDonald's® was criticized for using paperboard food containers by people concerned about the amount of trees being cut down to make these containers. So McDonald's switched to polystyrene foam "clamshells." Problem solved, right?

Wrong. People then criticized McDonald's for using polystyrene because they believed it generated a lot of waste, released chlorofluorocarbons (CFCs) that deplete the ozone layer, and it didn't decompose in landfills. Even though many of the claims of their opponents were untrue, McDonald's tried to satisfy the public by using non-CFC polystyrene and by developing recycling programs before switching back to paperboard in 1990.

But are paperboard containers more environmentally friendly? After performing life cycle analyses on paperboard and polystyrene clamshells, the Franklin Associates found that paperboard containers have some negative impacts on the environment, just as polystyrene containers do. The question of which is worse depends on personal perspective.

Myth: America faces a garbage crisis.

Fact: This myth could have some truth, depending upon how you define "crisis" and what is important to you. You may think that landfilling materials instead of recycling is proof of a crisis. Recovering resources may be more important to you than any savings or decreased environmental impact of landfilling. Any decision not to recycle means a loss of resources, which to you may constitute a crisis.

To prevent a waste crisis from developing, solid waste and manufacturing industries are working to develop new waste disposal and resource recovery methods and thinner and lighter materials that use less resources up front and can be easily recycled. What can the American public do to help prevent a crisis? The answer is up to you.

Think About It

1. *Do you believe America faces a garbage crisis? Why? How do you define "crisis"?*

2. *What are some other commonly held notions about garbage and solid waste? How could you determine whether these are myths or facts?*

3. *Is there anything that you once believed about waste management that you are now reconsidering? How could you get more information to help you make an informed decision?*

● References

Lesson 1: Introduction to Solid Waste

"Trash Trouble in Tacktown, Part A"

"Simulation: Crisis in Centre City"; Middletown Clean Community: Middletown, OH, 1993.

Kaufman, D.; Franz, C. *Biosphere 2000: Protecting Our Global Environment;* Kendall/Hunt, Dubuque, IA, 1995.

Student Information for "Risky Business"

Crouch, E.A.C.; Wilson, R. *Risk/Benefit Analysis;* Ballinger: Cambridge, MA, 1982.

Site Selection Process, Phase 4A: Selection of a Preferred Site (S) Generic Risk Assessment, 1, Final Report; ENVIRON Corporation: Washington DC, December 6, 1985.

Travis, C.C.; Cook, S.C. *Hazardous Waste Incineration and Human Health;* CRC: Boca Raton, FL, 1989.

The Garbage Gazette "Getting a Degree in Garbage"

Environmental Careers Organization Web Site. Careers. Available Positions. http://www.eco.org/career/body01_1.htm (accessed 30 July 1998).

National Association of Environmental Professionals Web Site. Information. WWW Resources. Environmental Careers. http;//www.naep.org./Internter/Resource1.html#CAREERS (accessed 30 July 1998).

SWANA Membership Brochure. Solid Waste Association of North America: Silver Spring, MD, 1998.

Lesson 2: Waste Characterization

Teacher Background on Waste Characterization

Alexander, J.H. *In Defense of Garbage;* Praeger: Westport, CT, 1993.

"Characterization of Municipal Solid Waste," U.S. Environmental Protection Agency, 1996.

"Municipal Solid Waste Landfill Survey," U.S. Environmental Protection Agency, 1986.

Wirka, J. *Wrapped in Plastics: The Environmental Case for Reducing Plastics Packaging;* Environmental Action Foundation: Washington, DC, 1988.

Teacher Notes for "What's Waste?"

EDT 586 "Teaching Environmental Education" course materials, Miami University, Oxford, OH, Spring 1993.

U.S. Environmental Protection Agency.

The Garbage Gazette "Garbage Takes a Vacation"

Hogan, B. "All Baled Up and No Place to Go," *New York State Conservationist.* 1988, *42* (4), 36–39.

Jones, F.R. "Fire Situation Pieces: Islip, New York," *EPA Journal,* 1989, *15* (2), 40.

The Garbage Gazette "Archaeologists Find Garbage Tells the Truth"

Rathje, W.; Murphy, C. *Rubbish: The Archaeology of Garbage.* HarperCollins: New York, 1992.

Lesson 3: Health Concerns

Teacher Background on Health Concerns

Bollet, A.J. *Plagues and Poxes: The Rise and Fall of Epidemic Disease;* Demos: New York, 1987.

Busvine, J.R. *Insects, Hygiene, and History;* Athlone: London, 1976.

Cloudsley-Thompson, J.L. *Insects and History;* St. Martin's: New York, 1976.

McNeill, W.H. *Plagues and Peoples;* Anchor/Doubleday: Garden City, NY, 1976.

Teacher Notes for "You *Can* Catch Me, You Dirty Rat"

Epidemics and Ideas; Ranger, T., Slack, P. Eds.; Cambridge University: Cambridge, England, 1992.

LaBerge, A.E.F. *Mission and Method: The Early Nineteenth Century French Public Health Movement;* Cambridge University: Cambridge, England, 1992.

"Medicine: MAJOR MEDICAL INSTITUTIONS: Public Health Services: HISTORY OF PUBLIC HEALTH: Developments from 1875." *Britannica Online.*
http://www.eb.com:180/cgi-bin/g?DocF=macro/5004/10/1000.html (accessed 24 April 1998).

Nikiforuk, A. *The Fourth Horseman;* M. Evans: New York, 1991.

Stuller, J. "Cleanliness has only recently become a virtue," *Smithsonian.* 1991, *21* (11), 126–135.

Teacher Notes for "Name That Disease"

Beveridge, W.I.B. *Influenza: The Last Great Plague;* Prodist: New York, 1977.

Biddle, W. *A Field Guide to Germs;* Henry Holt: New York, 1995.

Black, J.G. *Microbiology Principles and Applications,* 2nd ed.; Prentice Hall: Englewood Cliffs, NJ, 1993.

Bollet, A.J. *Plagues and Poxes: The Rise and Fall of Epidemic Disease;* Demos: New York, 1987.

Busvine, J.R. *Disease Transmission by Insects;* Springer-Verlag: Berlin, 1993.

Busvine, J.R. *Insects, Hygiene, and History;* Athlone: London, 1976.

Cloudsley-Thompson, J.L. *Insects and History;* St. Martin's: New York, 1976.

Color Atlas and Textbook of Diagnostic Microbiology, 5th ed.; Koneman, E.W., Allen, S.D., Janda, W.M., Schreckenberger, P.C., Winn, W.C., Eds; Lippincott: Philadelphia, 1997.

Dimmock, N.J.; Primrose, S.B. *Introduction to Modern Virology,* 3rd ed.; Blackwell Scientific: Oxford, 1987.

Garrett, L. *The Coming Plague;* Penguin: New York, 1994.

de Regnier, D.P.; Cook, S.A. Personal communications. Ferris State University, Big Rapids, MI, July, 1998.

The Garbage Gazette "A Thousand Years Without a Bath"

"Archimedes" *Britannica Online.* http://www.eb.com:180/cgi-bin/g?DocF=macro/5000/26.html (accessed 14 Aug 1998).

Black, J.G. *Microbiology Principles and Applications,* 2nd ed.; Prentice Hall: Englewood Cliffs, NJ, 1993.

Busvine, J.R. *Insects, Hygiene, and History;* Athlone: London, 1976.

Epidemics and Ideas; Ranger, T., Slack, P., Eds.; Cambridge University: Cambridge, England, 1992.

"Medicine: MAJOR MEDICAL INSTITUTIONS: Public health services: HISTORY OF PUBLIC HEALTH" *Britannica Online.* http://www.eb.com:180/cgi-bin/g?DocF=macro/5004/10/96.html (accessed 24 April 1998).

Stuller, J. "Cleanliness has only recently become a virtue," *Smithsonian.* 1991, *21* (11), 126–135.

The Garbage Gazette "Toxic Waste and You"

"Cytogenetic Patterns in Persons Living Near Love Canal—New York." *Morbidity and Mortality Weekly Report,* 1983, *32* (20), 261–262.

Fumento, M. *Science Under Siege;* Quill: New York, 1993.

Gorrie, P. "Bargain Prices Lure Buyers to Love Canal." *Toronto Star,* June 15, 1991, p A1.

Hoffman, A.J. "An Uneasy Birth at Love Canal." *Environment,* 1995, *37* (2), 4–9+.

Wildavsky, A. *But Is It True? A Citizen's Guide to Environmental Health and Safety Issues;* Harvard University: Cambridge, MA, 1995.

Lesson 4: Source Reduction

Teacher Background for Source Reduction

Denison, R.A.; Ruston, J. *Recycling and Incineration: Evaluating the Choices;* Island: Washington, DC, 1990.

Teacher Notes for "One Liter —To Go"

Heimlich, J.E. *Waste Wise: Concepts in Waste Management;* The Aseptic Packaging Council: Washington, DC, 1991.

Williams, S. "Trash Into Cash"; Investor Responsibility Research Center: Washington, DC, 1991.

Teacher Notes for "Wrap It Up"

All "Trashed" Out; Illinois Department of Energy and Natural Resources Office of Recycling and Waste Reduction, 1991, pp 22–24.

Pence, M. Personal communication; Woodland Elementary, Middletown, OH, 1993.

Selke, S.E.M. *Packaging and the Environment: Alternatives, Trends, and Solutions;* Technomic: Lancaster, PA, 1990.

Washington State Department of Ecology Litter Control and Recycling Program, 1985, pp 34–35.

The Garbage Gazette "War of the Packing Fillers"

Eco-Foam Product Information. American Excelsior Company: Arlington, TX: 1997.
Eco-Foam Web Site. http://www.eco-foam.com (accessed 14 Aug 1998).
"Loose-fill an Environmentalist Can Love," *Packaging Digest.* 1991, *28* (4), 44, 46.
"Plastic Foam Loose-Fill and the Environment, Plastic Loose-Fill Producer's Council: Grand
    Rapids, MI, 1994.
Polystyrene Packaging Council Web Site. http;//www.polystyrene.org (accessed 14 Aug 1998).

The Garbage Gazette "Enjoying Fresh Milk in the Desert?"

Aseptic Packaging Council Web Site. http://www.aseptic.org (accessed 14 Aug 1998).
Hayton, B. "50 Years of Food Innovations: the Hot, the Dry, and the Frozen," *Current Health 2.*
    1990: *17* (3), 17–19.
Hunter, B.T. "The Pros and Cons of Aseptic Packaging," *Consumer's Research Magazine.* 1991,
    74 (8), 15–17.
Insitute of Food Technologists Web Site. Education and careers. Introduction to the Food
    Industry. Lesson 1: Food Safety and Quality Assurance. http://www.ift.org/car/food_ind/
    mod2/html (accessed 14 Aug 1998).
Pardue, L. "Pandora's Juice Box? Jury Is Still Out on Aseptic Containers." *E Magazine.* 1992,
    3 (4), 50–52.
Raymond, M. "Aseptic Recycling on Trial," *BioCycle.* 1992, *33* (1), 80–84.
Steuteville, R. "Recycling Polycoated Packaging," *BioCycle.* 1994, *35* (3), 71–73.
"Ups and Downs of Aseptic," *BioCycle.* 1994, *35* (3), 74–54.

Lesson 5: Reusing Materials

Teacher Background on Reusing Materials

Duston, T.E. *Recycling Solid Waste: The First Choice for Private and Public Sector Management;*
    Quorum: Westport, CT, 1993.

Teacher Notes for "Treasures from Trash"

Wisconsin Department of Natural Resources, Bureau of Solid Waste Information and Education,
    Madison, WI.
Williams, S. "Trash Into Cash"; Investor Responsibility Research Center: Washington, DC, 1991.

Teacher Notes for "Shrinking Crafts"

"Facts and Figures of the U.S. Plastics Industry"; The Society of the Plastics Industry, 1992.
"Plastics in Perspective: Answers to Your Questions About Plastics in the Environment";
    American Plastics Council/The Society of Plastics Industry, 1993.

The Garbage Gazette "Earthships Take Off"

Climo, D. "The Home with Fewer Bills," *Building Products News.* 1997, *32* (7), 34–35.

"Focus Fact Sheet: Tire Recovery," Keep America Beautiful, Inc., Web Site. KAB Catalog. Focus Fact Sheet: Tire Recovery. http://www.kab.org/old/tires.html (accessed 28 May 1998).

Reed, S.; Haederle, M. "Want an Ecologically Correct House?," *PeopleWeekly.* 1991, *35* (1), 105–111.

"Scrap Tire Homes Keep Rolling Along," *BioCycle.* 1995, *36* (11), 58–59.

The Garbage Gazette "Don't Dismiss Disposables"

Fagin, D. "Down to a Science," *Newsday*, April 20, 1993, p 57.

Rathje, W.; Murphy, C. *Rubbish: The Archaeology of Garbage;* HarperCollins: New York, 1992.

Waste Policy Center; Perchard's. *Environmental and Public Health Aspects of Reusable and Disposable Food Service Packaging.* Waste Policy Center: Leesburg, VA, 1996.

Lesson 6: Resource Recovery

Teacher Background for "Now Separate It!" and "Trash in the Newspaper"

Denison, R.A.; Ruston, J. *Recycling and Incineration: Evaluating the Choices;* Island: Washington, DC, 1990.

Franklin Associates. *Characterization of Municipal Solid Waste in the United States.* Technical Report. Franklin Associates: Prairie Village, KS, 1996.

Earth Programs in Action Education Committee.

Ehrle, C. "Trash Into Cash," *Resources: The Magazine of Environmental Management;* Environmental Resources Management Group, *16* (1), February 1994.

Environmental Systems of America. Environmental Factoids Page. Materials. http://envirosystemsinc.com/factoids.html (accessed 9 July 1998).

Environmental Protection Agency Web Site. Student and Teachers. Teachers Lounge. Facts About the Environment. Municipal Solid Waste Factbook. Getting Started. Combustion. http://www.epa.gov/epaoswer/non-hw/muncpl/factbook/internet/comf/combust.htm#top (accessed 18 Aug 1998).

Selke, S.E.M. *Packaging and the Environment;* Technomic: Lancaster, PA, 1990.

Williams, S. "Trash Into Cash"; Investor Responsibility Research Center: Washington, DC, 1991.

Student Background for "Now Separate It!" and "Trash in the Newspaper"

Pearce, F. "Burn Me." *New Scientist.* 22 November 1997. Available online: http://www.newscientist.com/ns/971122/features.html (accessed 16 Aug 1998).

Porter, J.W. *Recycling in America...The 25% Solution.* Leesburg, Virginia: Waste Policy Center, 1996.

Tierney, J. "Recycling Is Garbage." *The New York Times Magazine*, June 30, 1996. Available online: http://researcher.sirs.com/cgi-bin/res-article-display?7PL124A+biodegradable+ (accessed 16 Aug 1998).

Teacher Notes for "Now Separate It!"

Selinger, B. *Chemistry in the Marketplace;* Harcourt Brace Jovanovich: Sydney, Australia, 1989.

Super Saver Investigators; Ohio Department of Natural Resources, Division of Litter Prevention and Recycling; 1992, pp 103–105.

Van Natta, S. Personal communication, White Oak Junior High School, Cincinnati, OH, 1993.

Teacher Notes for "Trash in the Newspaper"

Kids C.A.R.E.; Mead Corporation, Dayton, OH, 1991.

Wonder Science; American Chemical Society/American Institute of Physics: Washington, DC, November 1990.

Van Natta, S. Personal communication, White Oak Junior High School, Cincinnati, OH, 1993.

Teacher Background for "Compost Columns"

Denison, R.A.; Ruston, J. *Recycling and Incineration: Evaluating the Choices;* Island: Washington, DC, 1990.

"Municipal Solid Waste Landfill Survey"; U.S. Environmental Protection Agency, 1986.

Williams, S. "Trash Into Cash"; Investor Responsibility Research Center: Washington, DC, 1991.

Teacher Notes for "Compost Columns"

Alexander, J.H. *In Defense of Garbage.* Praeger: Westport, CT, 1993, pp 140–143.

Bottle Biology Project; Department of Plant Pathology, University of Wisconsin: Madison, WI, 1991.

Garland, G.A.; Grist, T.A.; Green, R.E. "The Compost Story: From Soil Enrichment to Pollution Remediation." *Biocycle,* 1995, *36* (10), 53–56.

Naar, J. *Design for a Livable Planet;* Harper & Row: New York, 1990.

The New Zealand Institute for Crop & Food Research, Limited, Home Page. Composting—Waste is a Valuable Resource. http://www.crop.cri.nz/broadshe/compost.htm (accessed 30 June 1998).

NREL http://nrelinfo.nrel.gov/business/international/info-energy/supplies/msw-resources.html (accessed 28 July 1998).

Teacher Background for "How Good Is Your Fuel?"

Denison, R.A.; Ruston, J. *Recycling and Incineration: Evaluating the Choices;* Island: Washington, DC, 1990.

"The Environmental Impact of Municipal Solid Waste Incineration"; The Coalition on Resource Recovery and the Environment, The U.S. Conference of Mayors, 1989.

"Resource Recovery in North America"; National Solid Wastes Management Association, 1991.

Selke, S.E.M. *Packaging and the Environment;* Technomic: Lancaster, PA, 1990.

Travis, C.C.; Cook, S.C. *Hazardous Waste Incineration and Human Health;* CRC: Boca Raton, FL, 1989.

Teacher Notes for "How Good Is Your Fuel?"

American Chemical Society, *ChemCom Chemistry in the Community;* Kendall/Hunt: Dubuque, IA, 1988.

Lowder, S. Personal communication, Tahoe-Truckee High School, Truckee, CA, 1993.

Sherman, M. Personal communication, Ursuline Academy, St. Louis, MO, 1993.

Super Saver Investigators; Ohio Department of Natural Resources, Division of Litter Prevention and Recycling, 1992, p 126.

Student Instructions for "How Good Is Your Fuel?"

Environmental Protection Agency Web Site. Student and Teachers. Teachers Lounge. Facts About the Environment. Municipal Solid Waste Factbook. Getting Started. Combustion. http://www.epa.gov/epaoswer/non-hw/muncpl/factbook/internet/comf/combust.htm#top (accessed 18 Aug 1998).

The Garbage Gazette "Fluff Up a Milk Jug for a Good Night's Sleep?"

Recycler's World Web Sire. http://www.recycle. net (accessed 14 Aug, 1998).

We're Environmentally Friendly Web Site. Recycled fabrics. http://www.nyu.edu/pages/minetta/mercer/recycled.html (accessed 7 Aug 1998).

Woodward, L. *Polymers All Around You.* Terrific Science: Middletown, OH, 1992.

The Garbage Gazette "Nature's Garbage Disposal"

Appelhof, M. "Bins Enter the Schools," *BioCycle.* 1994, *35* (10), 66–67.

Conrad, P. "Worm Composters in School Programs," *BioCycle,* 1995, *36* (2), 91.

Edwards, C.A. "Historical Overview of Vermicomposting," *BioCycle.* 1995, *36* (6), 56–58.

Farrell, M. Teaching Children about Verimcomposting," *BioCycle.* 1997, *38* (6) 78–80.

Koerner, B.I. "It's All Business for Worms," *U.S. News and World Report.* 1997, *123* (11), 53.

Ruggle, D. "Young Composters Learn with Worms," *BioCycle,* 1995, *36* (8), 72–75.

"Sharing Lunch...with Worms," *Environment.* 1994, *36* (10), 24.

Lesson 7: Disposal Methods

Teacher Background for "Believe It Can Rot—Or Not"

Denison, R.A.; Ruston, J. *Recycling and Incineration: Evaluating the Choices;* Island: Washington, DC, 1990.

Selke, S.E.M. *Packaging and the Environment;* Technomic: Lancaster, PA, 1990.

Teacher Notes for "Believe It Can Rot—Or Not"

EDT 586 "Teaching Environmental Education" course materials, Miami University, Oxford, OH, Spring 1993.

Teacher Background for "Household Hazardous Waste"

"Clean Up! Buy Right! Be Safe!" Hamilton County Environmental Services: Cincinnati, OH.

Denison, R.A.; Ruston, J. *Recycling and Incineration: Evaluating the Choices;* Island: Washington, DC, 1990.

Duston, T. E. *Recycling Solid Waste: The First Choice for Private and Public Sector Management;* Quorum: Westport, CT, 1993.

"Environmental Update"; Reckitt & Colman: Montvale, NJ, 1996.

"Household Hazardous Waste Recycling Outlets"; Hamilton County Environmental Services, Cincinnati, OH.

"Household Product Management Wheel"; Environmental Hazards Management Institute: Durham, NH, 1997.

Kaufman, D.; Franz, C. *Biosphere 2000: Protecting Our Global Environment,* 2nd ed.; Kendall/Hunt: Dubuque, IA, 1996.

"Pollution and Solid Waste." *Investigating Solid Waste Issues,* Ohio Department of Natural Resources: Columbus, OH, 1996, pp B-77–B-97.

Travis, C.C.; Cook, S.C. *Hazardous Waste Incineration and Human Health;* CRC: Boca Raton, FL, 1989.

Teacher Notes for "The Bottom Line(r)"

French, D. "Super Saturday Science Session—Waste Management"; Miami University Middletown: Middletown, OH, 1992.

The Garbage Gazette "A Timeline of Trash Disposal"

Deetz, J. *In Small Things Forgotten: The Archaeology of Early American Life.* Anchor/Doubleday: Garden City, NY: 1977.

"Environmental Works: Solid Waste Management: HISTORICAL BACKGROUND." Britannica Online. http://www.eb.com:180/cgi-bin/g?DocF=macro/5007/42/26.html (accessed 1 April 1998).

Hering, R.; Greeley, S.A.; *Collection and Disposal of Municipal Refuse.* McGraw-Hill: New York, 1921.

Kelly, K. *Garbage: The History and Future of Garbage in America;* Saturday Review: New York, 1973.

Rathje, W.; Murphy C. *Rubbish: The Archaeology of Garbage.* HarperCollins: New York, 1992.

Lesson 8: Conclusion

Teacher Notes for "Life Cycle Assessment"

Franklin Associates. *Resources and Environmental Profile Analysis of Polyethylene and Unbleached Paper Grocery Sacks.* Technical Report. Franklin Associates: Prairie Village, KS, 1990.

Franklin Associates. *Resources and Environmental Profile Analysis of Foam Polystyrene and Bleached Paperboard Containers.* Technical Report. Franklin Associates: Prairie Village, KS, 1990.

Franklin Associates. *Resources and Environmental Profile Analysis of High-Density Polyethylene and Bleached Paperboard Gable Milk Containers.* Technical Report. Franklin Associates: Prairie Village, KS, 1991.

Ehrenfeld, J.R. "Designing Green Goods," *The World and I,* 1995, *10* (4), 216.

Life Cycle Assessment. Franklin Associates, http://www.fal.com/LCA/lca.html (accessed 12 Aug 1998).

Teacher Notes for "Trash Trouble in Tacktown, Part B"

"Simulation: Crisis in Centre City"; Middletown Clean Community: Middletown, OH, 1993.

Kaufman, D.; Franz, C. *Biosphere 2000: Protecting Our Global Environment;* Kendall/Hunt, Dubuque, IA, 1995.

The Garbage Gazette "Garbage Fact and Fiction"

Rathje, W.; Murphy, C. *Rubbish: The Archaeology of Garbage.* HarperCollins: New York, 1992.

About the Author

Terrific Science Press is a nonprofit publisher housed in the federally and state-funded Center for Chemical Education (Miami University Middletown in Ohio). At the Center, educators and scientists have worked together since the mid-1980s to provide professional development for teachers through innovative approaches to hands-on, minds-on science.